DPH PHYSICS SERIES

TEXT BOOK OF THERMODYNAMICS

By

D.K. Jha

DISCOVERY PUBLISHING HOUSE
NEW DELHI-110002

First Published-2004

ISBN 81-7141-910-0

Published by

DISCOVERY PUBLISHING HOUSE
4831/24, Ansari Road, Prahlad Street,
Darya Ganj, New Delhi-110002 (India)
Phone: 23279245 • Fax: 91-11-23253475
E-mail:dphtemp@indiatimes.com

Printed at:

Tarun Offset Printers,
Delhi–110 053

Preface

This book "Text Book of Thermodynamics" is primarily intended for students preparing for degree and honours students of various universities. Thermodynamics include a large number of topics. Since the present day students is some what pressed for time, the treatment has been kept short and direct. Only such historical and additional information has been given as may possibly interest the more serious type of students. An attempt has been made to make the language as simple as possible.

We hope this book will be found useful by the students and teachers in the various institution of India.

Author

Contents

Conductors, Lee's Method for Liquids, Spherical Shell Method (Radial Flow of Heat), Cylindrical Flow of Heat, Thermal Conductivity of Rubber, Thermal Conductivity of Glass, Heat How Through a Compound Wall, Accretion of Ice on Ponds, Wiedemann-Franz Law, Practical Applications of Conduction of Heat, Davy's Safety Lamp, Convection, Application of Convection, Central Heating System, Change of Pressure with Height, Convectivc Equilibrium of the Atmosphere, Differential Air Thermoscope, Thermopile Properties of Thermal Radiation, Applications of Heat Radiations, Thermos Flask, Black Body, Kirchoff 's Law of Heat Radiation, Prevost Theory of Heat Exchange, Lummer and Kurlbaum Bolometer,

Thermodynamics System

1.1 THERMODYNAMIC SYSTEM

A thermodynamic system is one which can be described in terms of the thermodynamic co-ordinates. The co-ordinates of a thermodynamic system can be specified by any pair of quantities *viz.*, pressure (P), volume (7), temperature (T) and entropy (S). The thermodynamic systems in engineering are gas, vapour, steam, mixture of gasoline vapour and air, ammonia vapours and its liquid. In Physics, thermodynamics includes besides the above, systems like stretched wires, thermocouples, magnetic materials, electrical condenser, electrical cells, solids and surface films.

1. Stretched Wire

In a stretched wire, to find the Young's modulus of a wire by stretching, the complete thermodynamic co-ordinates are

(a) the stretching force F,

(b) the length of the stretching wire, and

(c) the temperature of the wire.

The pressure and volume are considered to be constant.

2. Surface Films

For liquid films, in the study of surface tension, the thermodynamic co-ordinates are:

(a) the surface tension,

(b) the area of the film, and

(c) the temperature.

3. Reversible Cell

The thermodynamic coordinates to completely describe a reversible cell are:

(a) the E.M.F. of the cell,

(b) the charge that flows, and

(c) the temperature.

1.2 THERMAL EQUILIBRIUM AND CONCEPT OF TEMPERATURE (ZEROTH LAW OF THERMODYNAMICS)

A thermodynamic system is said to be in thermal equilibrium if any two of its independent thermodynamic co-ordinates X and Y remain constant as long as the external conditions remain unaltered. Consider a gas enclosed in a cylinder fitted with a piston. If the pressure and volume of the enclosed mass of gas are P and V at the temperature of the surroundings, these values of P and V will remain constant as long as the external conditions *viz*., temperature and pressure remain unaltered. The gas is said to be in thermal equilibrium with the surroundings.

The *zeroth law* of thermodynamics was formulated after the first and the second laws of thermodynamics have been enunciated. This law helps to define the term *temperature* of a system.

This law states that *if, of three systems. A, B and C, A and B are separately in thermal equilibrium with C, then A and B are also in thermal equilibrium with one another.*

Conversely the law can be stated as follows :

If three or more systems are in thermal contact, each to each, by means of diathermal walls and are all in thermal equilibrium together, then any two systems taken separately are in thermal equilibrium with one another.

Consider three fluids A, B and C. Let P_A, F_A represent the pressure and volume of A, P_B, V_B, the pressure and volume of B, and P_C, V_C are the pressure and volume of C.

If A and B are in thermal equilibrium, then

$$\phi_1(P_A, V_A) = \phi_2(P_B, V_B)$$

or $$F_1[P_A, V_A, P_B, V_B] = 0 \qquad \text{...(i)}$$

Expression (i) can be solved, and

$$P_B = Af_1[P_A, V_A, V_B] \qquad \text{...(ii)}$$

If B and G are in thermal equilibrium

$$\phi_2(P_B, V_B) = \phi_3(P_C, V_C)$$

or $$F_2[P_B, V_B, P_C, V_C] = 0$$

Also $$P_B = f_2[V_B, P_C, V_C] \qquad \text{...(iii)}$$

From equations (ii) and (iii) for A and C to be in thermal equilibrium separately,

$$f_1(P_A, V_A, V_B) = f_2[V_B, P_C, V_C). \qquad \text{...(iv)}$$

If A and C are in thermal equilibrium with B separately, then according to the zeroth law, A and C are also in thermal equilibrium with one another.

$$\therefore \quad F_3[P_A, V_A, P_C, V_C] = 0. \qquad \text{...(v)}$$

Equation (iv) contains a variable V_B, whereas equation (v) does not contain the variable V_B. It means

$$\phi_1(P_A, V_A) = \varphi_3(P_C, V_C) \qquad \text{...(vi)}$$

In general,

$$\phi_1(P_A, V_A) = \phi_2(P_B, V_B) = \phi_3(P_C, V_C) \qquad \text{...(vii)}$$

These three functions have the same numerical value though the parameters (P, V) of each are different. This numerical value is termed as *temperature* (T) of the body.

$$\therefore \quad \phi(P, V) = T \qquad \text{...(viii)}$$

This is called the equation of state of the fluid.

Therefore, the temperature of a system can be defined as the property that determines whether or not the body is in thermal equilibrium with the neighbouring systems. If a number of systems are in thermal equilibrium, this common property of the system can be represented by a single numerical value called the temperature. It means that if two systems are not in thermal equilibrium, they are at different temperatures.

In a mercury in glass thermometer, the pressure above the mercury column is zero and volume of mercury measures the temperature. If a thermometer shows a constant reading in two systems A and B separately, it will show the same reading even when A and B are brought in contact.

1.3 CONCEPT OF HEAT

Heat is defined as *energy in, transit*. As it is not possible to speak of work in a body, it is also not possible to speak of heat in a body. Work is either done on a body or by a body. Similarly, heat can flow from a body or to a body. If a body is at a constant temperature, it has both mechanical and thermal energies due to the molecular agitations and it is not possible to separate them. So, in this case, we cannot talk of heat energy. It means, if the flow of heat stops, the word heat cannot be used. It is only used when there is transfer of energy between two or more systems.

Consider two systems A and B in thermal contact with one another and surrounded by adiabatic walls.

For the system A,

$$H = U_2 = U_1 + W, \qquad \text{...(i)}$$

where B is the heat energy transferred, U_1 is the initial internal energy, U_2 is the final internal energy and W is the work done.

Similarly for the system B,

$$H' = U_2' = U_1' + W' \qquad \text{...(ii)}$$

Adding (i) and (ii)

$$\therefore \quad H + H' = (U_2 - U_1) + W + (U_2' = U_1') + W'$$

$$H + H' = [(U_2 + U_2') - (U_1 + U_1')] + (W + W') \qquad \text{...(iii)}$$

The total change in the internal energy of the composite system

$$= [(U_2 + U_2') - (U_1 + U_1')]$$

The work done by the composite system = W + W'

It means that the heat transferred by the composite system = H + H'. But the composite system is surrounded by adiabatic and the net heat transferred is zero.

$$\therefore \quad H + H' = 0$$

$$H = -H' \qquad \text{...(iv)}$$

Thus, for two systems A and B in thermal contact with each and other the composite system surrounded by adiabatic walls, the heat gained by one system is equal to the heat lost by the other system.

1.4 QUASISTATIC PROCESS

A system in thermodynamical equilibrium must satisfy the ring requirements strictly:

(i) Mechanical Equilibrium

For a system to be in mechanical equilibrium, there should be no unbalanced forces acting on any part of the system or the system as a whole.

(ii) Thermal Equilibrium

For a system to be in thermal equilibrium, there should be no temperature difference between the parts of the system or between the system and the surroundings.

(iii) Chemical Equilibrium

For a system to be in chemical equilibrium, there should be no chemical reaction within the system and also no movement of any chemical constituent from one part of the system to the other.

When a system is in thermodynamic equilibrium and the surroundings are kept unchanged, there will be no motion and also no work will be done. On the other hand, if the sum of the external is altered, resulting in a finite unbalanced force acting on the system, the condition for mechanical equilibrium will not be satisfied any longer. This results in the following:

(i) Due to unbalanced forces within the system, turbulence, waves etc. may be set up. The system as a whole may possess an accelerated motion.

(ii) Due to turbulence, acceleration etc. the temperature distribution within the system may not be uniform. There may also exits a finite temperature difference between the system and the surroundings.

(iii) Due to the presence of unbalanced forces and difference in temperature, chemical reaction may take place or there may be movement of a chemical constituent.

From this discussion, it is clear that a finite unbalanced force may cause the system to pass through non-equilibrium states. If during a thermodynamic process, it is desired to describe every state of a system by thermodynamic coordinates referred to the system as a whole, the process should not be brought about by a finite unbalanced force.

A quasistatic process is defined as the process in which the deviation from thermodynamic equilibrium is infinitesimal and all the states through which the system passes during a quasistatic process can be considered as equilibrium states.

In actual practice, many processes closely approach a quasistatic process and may be treated as such with no significant error. Consider the expansion of a gas in a closed cylinder fitted with a piston. Initially weights are on the piston and the pressure of the gas inside the cylinder is higher than the atmospheric pressure. If the weights are small and are taken off slowly one by one, the process can be considered quasistatic. If, however, all the weights are removed at once, expansion takes place suddenly and it will be a nonequilibrium process. The system will not be in equilibrium at any time during this process.

A quasistatic process is an ideal concept that is applicable to all thermodynamic systems including electric and magnetic systems. It should be noted that conditions for such a process can never be satisfied rigorously in practice.

1.5 HEAT—PATH FUNCTION

Heat is a path function. When a system changes from a state 1 to state 2, the quantity of heat transferred will depend upon the intermediate stages through which the system passes *i.e.*, its path. Hence heat is an Inexact *differential* and is written as δB.

On integrating, we get

$$\int_{1A}^{2A} \delta H = H_{1A}^{2A}$$

Here, ${}_1H_2$, represents the heat transferred during the process between the states 1 and 2 along a particular path A.

1.6 WORK—A PATH FUNCTION

Suppose that a system is taken from an initial equilibrium state 1 to a final equilibrium state 2 by two different paths A and B (Fig. 1.1). The processes are quasistatic.

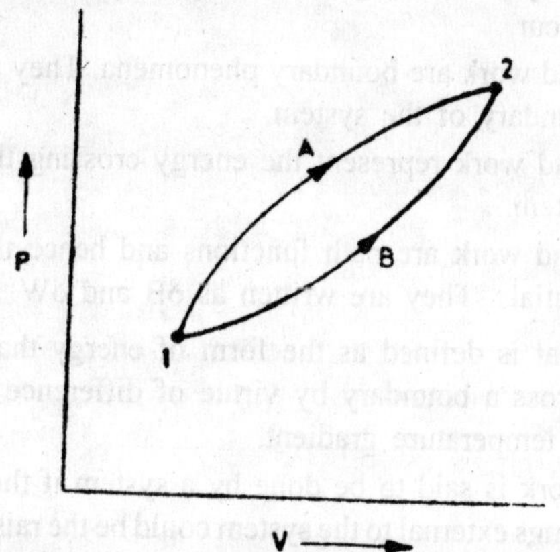

Fig. 1.1

The areas under these curves are different and hence the quantities of work done are also different.

For the path A, $W_A = \int_{1A}^{2A} \delta W = \int_{1A}^{2A} P\,dV$ ...(i)

For the path B $W_B = \int_{1B}^{2B} \delta W = \int_{1B}^{2B} P\,dV$ ...(ii)

The values of W_A and W_B are not equal. Therefore work cannot be expressed as a difference between the values of some property of the system in the two states. Therefore, it is not correct to represent

$$W = \int_{W_1}^{W_2} \delta W = W_2 - W_1. \quad ...(iii)$$

It may be pointed out that it is meaningless to say "work in a system or work of a system". Work cannot be interpreted similar to temperature or pressure of a system.

In terms of calculus δW is an inexact differential. It means that W is not a property of the system and $\int \delta W$ cannot be expressed as the difference between two quantities that depend entirely on the initial and the final states.

Hence, heat and work are *path functions* and they depend only on the process They are not point functions such as pressure or temperature. Work done in taking the system from state 1 to state 2 will be different for different paths.

1.7 COMPARISON OF HEAT AND WORK

There are many similarities between heat and work. These are:

1. Heat and work are both transient phenomena. Systems do not possess heat or work.

2. When a system undergoes a change, heat transfer or work done may occur.
3. Heat and work are boundary phenomena. They are observed at the boundary of the system.
4. Heat and work represent the energy crossing the boundary of the system.
5. Heat and work are path functions and hence they are inexact differentials. They are written as δB and δW.
6. (a) Heat is defined as the form of energy that is transferred across a boundary by virtue of difference of temperature or temperature gradient.

   (b) Work is said to be done by a system if the sole effect on things external to the system could be the raising of a weight.

It is customary to represent, work done by the system as +ve, work done on the system as –ve, heat flowing into the system as +ve, and heat flowing out of the system as –ve.

1.8 FIRST LAW OF THERMODYNAMICS

Joule's law gives the relation between the work done and the heat produced. It is true when the whole of the work done is used in producing heat or vice versa. Here, $W = JH$ where J is the Joule's mechanical equivalent of heat. But in practice, when a certain quantity of heat is supplied to a system the whole of the heat energy may not be converted into work. Part of the heat may be used in doing external work and the rest of the heat might be used in increasing the internal energy of the molecules. Let the quantity of heat supplied to a system be δH, the amount of external work done be δW and the increase in internal energy of the molecules be dU. The term U represents the internal energy of a gas due to molecular agitation as well as due to the forces of intermolecular attraction. Mathematically

$$\delta H = dU + \delta W \qquad \text{...(i)}$$

Equation (i) represents the first law of thermodynamics. All the quantities are measured in heat units. The first law of thermodynamics states that the amount of heat given to a system is equal to the sum of the increase in the internal energy of the system and the external work done.

For a cyclic process, the change in the internal energy of the system is zero because the system is brought back to the original condition. Therefore for a cyclic process $\oint \delta U = 0$

and $$\oint \delta H = \oint dw \qquad \text{...(ii)}$$

[Both are expressed in heat units].

This equation represents Joule's law.

For a system carried through a cyclic process, its initial and final internal energies are equal. From the first law of thermodynamics, for a system undergoing any number of complete cycles

$$U_2 - U_1 = 0$$

$$\oint \delta H = \oint \delta W$$

$$H = W \text{ [Both are in heat units].}$$

1.9 FIRST LAW OF THERMODYNAMICS FOR A CHANGE IN STATE OF A CLOSED SYSTEM

For a closed system during a complete cycle, the first law of thermodynamics is written as

$$\oint \delta H = \oint \delta W$$

In practice, however, we are also concerned with a process rather than a cycle. Let the system undergo a cycle, changing its state from 1 to 2 along the path A and from 2 to 1 along the path B. This cyclic process is represented in the P – V diagram (Fig. 1.2).

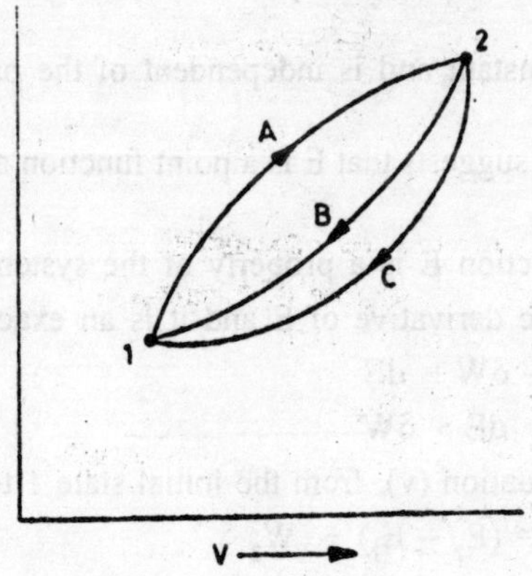

Fig. 1.2

According to the first law of thermodynamics

$$\oint \delta H = \oint \delta W$$

For the complete cyclic process

$$\int_{1A}^{2A} \delta H + \int_{2B}^{1B} \delta H = \int_{1A}^{2A} \delta W + \int_{2B}^{1B} \delta W \quad ...(i)$$

Now, consider the second cycle in which the system changes from state I to state 2 along the path A and returns from state 2 to state 1 along the path C. For this cyclic process

$$\int_{1A}^{2A} \delta H + \int_{2C}^{1C} \delta H = \int_{1A}^{2A} \delta W + \int_{2C}^{1C} \delta W \quad ...(ii)$$

Subtracting (ii) from (i)

$$\int_{2B}^{1B} \delta H - \int_{2C}^{1C} \delta H = \int_{2B}^{1B} \delta W - \int_{2C}^{1C} \delta W$$

or $$\int_{2B}^{1B} (\delta H - \delta W) = \int_{2C}^{1C} (\delta H - \delta W) \quad ...(ii)$$

Here B and C represent arbitrary processes between the states 1 and 2. Therefore, it can be concluded that the quantity $(\delta H - \delta W)$ is the same for all processes between the states 1 and 2. The quantity $(\delta H - \delta W)$ depends only on the initial and the final states of the system and is independent of the path followed between the two states.

Let $dE = (\delta H - \delta W)$

From the above logic, it can be seen that

$$\int_1^2 dE = \text{constant and is independent of the path.}$$

This naturally suggests that E is a point function and dE is an exact differential.

The point function E is a property of the system.

Here dE is the derivative of E and it is an exact differential.

$$\therefore \quad \delta H - \delta W = dE \quad ...(iv)$$

or $$\delta H = dE + \delta W \quad ...\{v)$$

Integrating equation (v), from the initial state 1 to the final state 2

$${}_1H_2 = (E_2 - E_1) + {}_1W_2$$

Note: ${}_1H_2$ cannot be written as $(H_2 - H_1)$, because it depends upon the path.

Similarly, ${}_1W_2$ cannot be written as $(W_2 - W_1)$, because it also depends upon the path.

Here ${}_1H_2$ represents the heat transferred,

${}_1W_2$ represents the work done,

E_2 represents the total energy of the system in state 2,

E_1 represents the total energy of the system in state 1.

At this point, it is worthwhile discussing what this E can possibly mean. With reference to the system, the energies crossing the boundaries are all taken care of in the form of B and W. For dimensional stability of Eq. (v), this E must be energy and this must belong to the system. Therefore,

E_2 represents the energy of the system in state 2

E_1 represents the energy of the system in state 1

This energy E acquires a value at any given equilibrium condition by virtue of its thermodynamic state. The working substance, for example a gas, has molecules moving in all random fashion. The molecules have energy associated by virtue of mutual attraction and this part is similar to the potential energy of a body in macroscopic terms. They also have velocities and hence kinetic energy. This energy E therefore can be visualised as comprising of molecular potential and kinetic energies in addition to macroscopic potential and kinetic energies, The first part, which owes its existence to the thermodynamic nature is often called the internal energy which is completely dependent on the thermodynamic state and the other two depend on mechanical or physical state of the system.

E = U + KE + PE + Others which depend upon chemical nature etc.

For a closed system (non-chemical) the changes in all others except U are insignificant and

$$dE = dU$$

∴ From equation (v)

$$\delta H = dU + \delta W \qquad \text{...(vi)}$$

Here all the quantities are in consistent units

1.10 APPLICATIONS OF FIRST LAW OF THERMODYNAMICS SPECIFIC HEAT OF A GAS (T AND V INDEPENDENT)

The internal energy of a system is a angle valued function of the state variables viz., pressure, volume, temperature etc. In the case of a gas, any two of the variables P, V, T are sufficient to define completely its state. If V and T are chosen as the independent variables,

$$U = f(V, T) \qquad \text{...(i)}$$

Differentiating equation (i)

$$dU = \left(\frac{\partial U}{\partial T}\right)_V dT + \left(\frac{\partial U}{\partial V}\right)_T dV \qquad \text{...(ii)}$$

If an amount of heat δH is supplied to a thermodynamical system, say an ideal gas and if the volume increases by dV at a constant pressure P, then according to the first law of thermodynamics

$$\delta H = dU + \delta W$$

Here $\delta W = P\,.\,dV$

$\therefore \quad \delta H = dU + P\,.\,dV$

Substituting the value of W from equation (ii)

$$\delta H = \left(\frac{\partial U}{\partial T}\right)_V dT + \left(\frac{\partial U}{\partial V}\right)_T dV + P.dV \qquad \text{...(iii)}$$

Dividing both sides by dT

$$\frac{\delta H}{dT} = \left(\frac{\partial U}{\partial T}\right)_V + \left(\frac{\partial U}{\partial V}\right)_T \frac{dV}{dT} + \frac{P.dV}{dT}$$

or

$$\left(\frac{\delta H}{dT}\right) = \left(\frac{\partial U}{\partial T}\right)_V + \left[P + \left(\frac{\partial U}{\partial V}\right)_T\right]\frac{dV}{dT} \qquad \text{...(iv)}$$

If the gas is heated at constant volume,

$$\left(\frac{\delta H}{\partial T}\right)_V = C_V \quad \text{and} \quad \frac{dV}{dT} = 0$$

$$\therefore \quad \left(\frac{\delta H}{\partial T}\right)_V = \left(\frac{\partial U}{\partial T}\right)_V = C_V \qquad \text{...(v)}$$

When the gas is heated at constant pressure,

$$\left(\frac{\delta H}{\partial T}\right)_P = C_P$$

$\therefore$ From equation (iv),

$$C_P = \left(\frac{\partial U}{\partial T}\right)_V + \left[P + \left(\frac{\partial U}{\partial V}\right)_T\right]\left(\frac{\partial V}{\partial T}\right)_P$$

$$\therefore \qquad C_P = C_V + \left[P + \left(\frac{\partial U}{\partial V}\right)_T\right]\left(\frac{\partial V}{\partial T}\right)_P$$

$$\text{or} \qquad C_P - C_V = \left[P + \left(\frac{\partial U}{\partial V}\right)_T\right]\left(\frac{\partial V}{\partial T}\right)_P \qquad \text{...(vi)}$$

From Joule's experiment, for an ideal gas on opening the stopcock, no work was done and no heat transfer tool; place.

So $\delta H = 0 = \delta U + 0$. Therefore, $dU = 0$. Even though the volume changed while the temperature is constant, there if no change in internal energy.

$$\left(\frac{\partial V}{\partial T}\right)_P = 0$$

From the ideal gas equation

$$PV = RT$$

$$\text{or} \qquad P\left(\frac{\partial V}{\partial T}\right)_P = R \qquad \text{...(vii)}$$

$$\therefore \qquad C_P - C_V = P\left(\frac{\partial V}{\partial T}\right)_P + \left(\frac{\partial U}{\partial V}\right)_T\left(\frac{\partial V}{\partial T}\right)_P$$

$$\text{But} \qquad \left(\frac{\partial U}{\partial V}\right)_T = 0$$

$$\therefore \qquad C_P - C_V = P\left(\frac{\partial V}{\partial T}\right)_P = R$$

$$\therefore \qquad C_P - C_V = R \qquad \text{...(viii)}$$

Here C_P, C_V and R are expressed in the same units.

From equation (iii)

$$\delta H = \left(\frac{\partial V}{\partial T}\right)_V dT + \left[P + \left(\frac{\partial U}{\partial V}\right)_T\right]dV \qquad \text{...(ix)}$$

For a process at constant temperature

$$dT = 0$$

$$\therefore \qquad (\delta H)_T = P(dV)_T + \left(\frac{\partial U}{\partial V}\right)_T (dV)_T \qquad \text{...(x)}$$

This equation represents the amount of heat energy supplied to a system in an isothermal reversible process and is equal to the sum of the work done by the system and the increase in its internal energy.

For a reversible adiabatic process

$$\delta H = 0,$$

Therefore, from equation (ix),

$$0 = C_V\, dT + \left[P + \left(\frac{\partial U}{\partial V}\right)_T\right] dV$$

or
$$C_V dT = -\left[P + \left(\frac{\partial U}{\partial V}\right)_T\right] dV$$

Dividing throughout by dV,

$$C_V\left(\frac{\partial T}{\partial V}\right) = -\left[P + \left(\frac{\partial U}{\partial V}\right)_T\right] \qquad \text{...(xii)}$$

The isobaric volume coefficient of expansion

$$\alpha = \frac{1}{V}\left(\frac{\partial V}{\partial T}\right)_P$$

$$\therefore \quad \alpha V = \left(\frac{\partial V}{\partial T}\right)_P \qquad C_P - C_V = P\left(\frac{\partial V}{\partial T}\right)_P$$

$$\therefore \quad \frac{C_P - C_V}{\alpha V} = P$$

But
$$\left(\frac{\partial U}{\partial V}\right)_T = 0 = P - P$$

or
$$\left(\frac{\partial U}{\partial V}\right)_T = \left(\frac{C_P - C_V}{\alpha V}\right) - P \qquad \text{...(xiii)}$$

or
$$-\left(\frac{C_P - C_V}{\alpha V}\right) = \left[P + \left(\frac{\partial U}{\partial V}\right)_T\right] \qquad \text{...(xiv)}$$

From equations (xii) and (xiv)

$$C_V\left(\frac{\partial T}{\partial V}\right) = -\left(\frac{C_P - C_V}{\alpha V}\right)$$

or
$$\left(\frac{\partial T}{\partial V}\right) = \frac{C_V - C_P}{\alpha V C_V} \qquad \text{...(xv)}$$

This expression holds good for an adiabatic reversible process.

1.11 ISOTHERMAL PROCESS

If a system is perfectly conducting to the surroundings and the temperaturc remains constant throughout the process, it is called an isothermal process. Consider a working substance at a certain pressure and temperature and having a volume represented by the point A (Fig. 1.3).

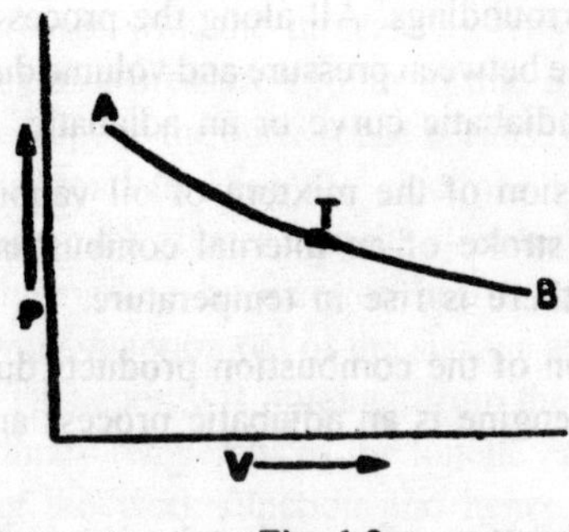

Fig. 1.3

Pressure is decreased and work is done by the working substance at the cost of its internal energy and there should be fall in temperature. But, toe system is perfectly conducting to the surroundings. It absorbs heat from the surroundings and maintains a constant temperature. Thus from A to B the temperature remains constant. The curve AB a called the *isothermal* curve or *isothermal*.

Consider the working substance at the point B and let the pressure be increased. External work is done on the working substance and there should be rise in temperature. But the system is perfectly conducting to the surroundings. It gives extra heat to the surroundings and its temperature remains constant from B to A.

Thus, during the isothermal process, the temperature of the working substance remains constant. It can absorb heat or give heat to the surroundings. The equation far an isothermal process is

$$PV = RT = \text{constant (For one gram molecule of a gas.)}$$

For n gram molecules of a gas $PV = nRT$

1.12 ADIABATIC PROCESS

During an adiabatic process, the working substance is perfectly insulated from the surroundings. It can neither give heat nor take heat from the surroundings. Work is done on the working substance, there is rise in temperature because the external work done on the working substance increases its internal energy. When work is done by the working substance, it is done at the cost of its internal energy. As the system is perfectly insulated from the surroundings, there is fall in temperature.

Thus, during an adiabatic process, the working substance is perfectly insulated from the surroundings. All along the process, there is change in temperature. A curve between pressure and volume during the adiabatic process is called an adiabatic curve or an adiabatic.

1. The compression of the mixture of oil vapour and air during compression stroke of an internal combustion is an adiabatic process and there is rise in temperature.
2. The expansion of the combustion products during the working stroke of an engine is an adiabatic process and there is fall in temperature.
3. The sudden bursting of a cycle tube is an adiabatic process.

Apply the first law of thermodynamics to an adiabatic process, $\delta H = 0$.

$$\delta H = dU + \delta W$$

or $$0 = dU + \delta W \quad \text{...(i)}$$

The processes that take place *suddenly* or *quickly* are adiabatic processes.

1.13 ISOCHORIC PROCESS

If the working substance is taken in a non-expanding chamber, the heat supplied will increase the pressure and temperature. The volume of the substance will remain constant. Such a process is called an *isochoric process*. The work done is zero because there is no change in volume. The whole of the heat supplied increased the internal energy. Therefore, during the isochoric process $\delta W = 0$.

$$\delta H = dU \quad \text{...(i)}$$

The heat transferred in such a process

$$\delta H = C_v dT$$

$$\therefore \quad C_v dT = dU \quad \text{...(ii)}$$

Hence C_v is the specific heat for one gram-molecule of a gas at constant volume.

1.14 ISOBARIC PROCESS

If the working substance is taken in an expanding chamber kept at a constant pressure, the process is called an isobaric process. Here, the temperature and volume change. If an amount of heat δB is given to the working substance, it is partly used in increasing the temperature of the working substance by dT and partly used in doing external work. Considering one gram of the working substance,

$$\delta H = 1 \times C_V dT + \frac{P.dV}{J} \quad \text{...(i)}$$

But $$\delta H = C_P.dT \quad \text{...(ii)}$$

$$P.dV = r.dT$$

$$\therefore \quad C_P dT = C_V.dT + \frac{r.dY}{J}$$

$$C_P - C_P = r/J \quad \text{...(iii)}$$

Here C_P and C_V represent the specific beats for 1 gram of a gas and r is the ordinary gas constant

If C_P and C_V are the gram-molecular specific heats of gas, then

$$C_P - C_V = \frac{R}{J} \quad \text{...(iv)}$$

Here A is the universal gas constant.

1.15 GAS EQUATION DURING AN ADIABATIC PROCESS

Consider 1 gram of the working substance (ideal gas) perfectly insulated from the surroundings. Let the external work done by the gas be δW.

Applying the first law of thermodynamics

$$\delta H = dU + \delta W$$

But $$\delta H = 0$$

and $\delta W = P.dV$

where P is the pressure of the gas and dV is the change in volume.

$$\therefore \quad 0 = dU + \frac{P.dV}{J}$$

As the external work Is done by the gas at the cost of its internal energy, there is fall in temperature by dT.

$$dU = 1 \times C_v \times dT$$

$$C_V.dT + \frac{P.dV}{J} = 0 \qquad ...(ii)$$

For an ideal gas

$$PV = rT \qquad ...(iii)$$

Differentiating,

$$P.dV + V.dP = r.dT$$

Substituting the value of dT in equation (ii),

$$C_V\left[\frac{P.dV + V.dP}{r}\right] + \frac{P.dV}{J} = 0$$

$$C_V[P.dV + V.dP] + r.\frac{P.dV}{J} = 0$$

But, $\frac{r}{J} = C_P - C_V$

$$\therefore C_v.P.dV + C_v.V.dP + C_p.PdV - C_vPdV = 0$$

$$C_p.P.dV + C_v.V.dP = 0$$

Integrating, $C_v.PV$,

$$\frac{C_P}{C_V}.\frac{dV}{V} + \frac{dP}{P} = 0$$

But $\frac{C_P}{C_V} = 0$

$$\therefore \quad \frac{dP}{P} + \gamma\frac{dV}{V} = 0$$

Integrating, $\log P + \gamma \log V = \text{const.}$

or $\log PV^{\gamma} = \text{const.}$

$$PV^{\gamma} = \text{const.} \qquad \text{...(iv)}$$

This is the equation connecting pressure and volume during an adiabatic process.

Taking $PV = rT$

or $P = \dfrac{rT}{V}$

$$\left(\frac{rT}{V}\right).V^{\gamma} = \text{cosnt.}$$

But r is const.

$$rTV^{\gamma-1} = \text{cosnt.}$$

$$\therefore \quad rTV^{\gamma-1} = \text{const.} \qquad \text{...(v)}$$

Also $V = \dfrac{rT}{P}$

$$P\left[\frac{rT}{P}\right]^{\gamma} = \text{cosnt.}$$

or $\dfrac{r^{\gamma}T^{\gamma}}{P^{\gamma-1}} = \text{const.}$

or $\dfrac{P^{\gamma-1}}{T^{\gamma}} = \text{const.}$ ...(vi)

Thus, during an adiabatic process

(i) $PV_{\gamma} = \text{const.}$

(ii) $TV^{\gamma-1} = \text{const.}$ and

(iii) $\dfrac{P^{\gamma-1}}{T^{\gamma}} = \text{const.}$

1.16 SLOPES OF ADIABATICS AND ISOTHERMALS

In an isothermal process

$$PV = \text{const.}$$

Differentiating,

$$PdV + VdP = 0$$

or $\dfrac{dP}{dV} = -\dfrac{P}{V}$ ...(i)

In an adiabatic process

$$PV^{\gamma} = \text{const.}$$

Differentiating,

$$P_{\gamma}V^{\gamma-1}\, dV + VdP = 0$$

$$\frac{dP}{dV} = -\frac{\gamma P}{V} \qquad \text{...(ii)}$$

Therefore, the slope of an adiabatic is γ times the slope of the isothermal.

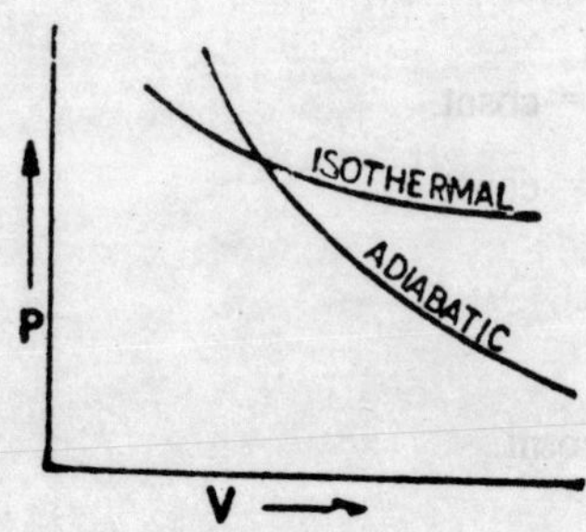

Fig. 1.4

Hence, the adiabatic curve is steeper than the isothermal curve (Fig. 1.4) at a point where the two curves intersect each other.

1.17 WORK DONE DURING AN ISOTHERMAL PROCESS

When a gas is allowed to expand isothermally, work is done by it. Let the initial and final volumes be V_1 and V_2 respectively. In Fig. 1.5, the area of the shaded strip represents the work done for a small change in volume dV. When the volume changes from V_1 to V_2,

$$\text{Work done} = \int_{V_1}^{V_2} P.dV = \text{area ABba} \qquad \text{...(i)}$$

Fig. 1.5 represents the indicator diagram. Considering one gram molecule of the gas

$$PV = BT$$

or $$P = \frac{RT}{V}$$

$$\therefore \quad W = RT \int_{V_1}^{V_2} \frac{dV}{V}$$

$$= RT \log_e \frac{V_2}{V_1} \quad ...(ii)$$

$$W = RT \times 2.3026$$

$$\log_{10} \frac{V_2}{V_3} \quad ...(iii)$$

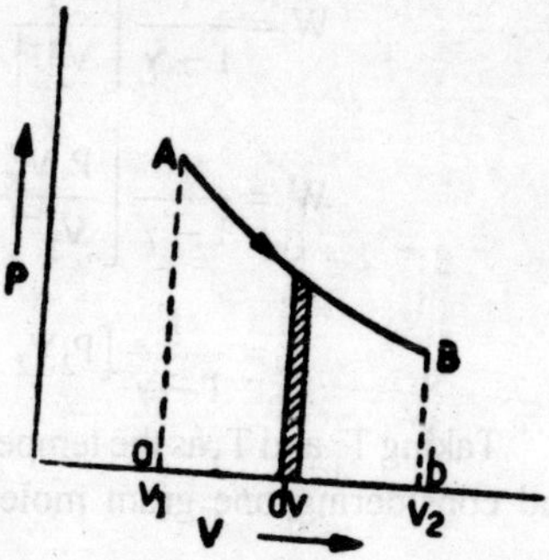

Also $P_1V_1 = P_2V_2$

or $$\frac{V_2}{V_1} = \frac{P_1}{P_2}$$

Fig. 1.5

$$\therefore W = RT \times 2.3026 \times \log_{10} \frac{P_1}{P_2} \quad ...(iv)$$

Here, the change in the internal energy of the system is zero (because the temperature remains constant). So the heat transferred is equal to the work done.

1.18 WORK DONE DURING AN ADIABATIC PROCESS

During an adiabatic process, the gas expands from volume V_1 to V_2. As shown by the indicator diagram (Fig. 1.6) the work done for an increase in volume dV = P.dV. Work done when the gas expands from V_1 to V_2 is given by,

$$W = \int_{V_1}^{V_2} PdV = \text{Area ABba}$$

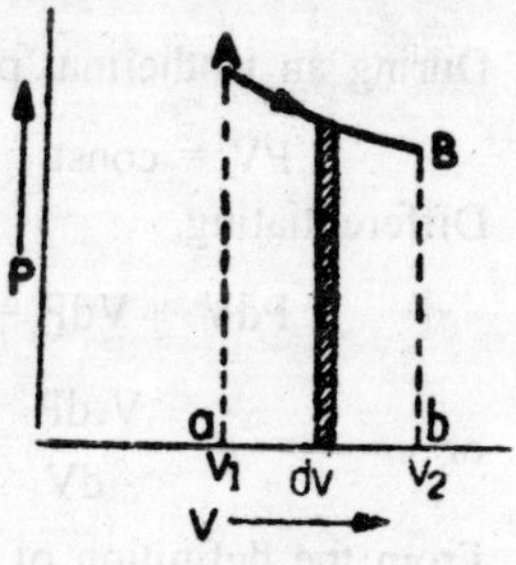

During an adiabatic process,

$$PV^\gamma = \text{cosnt} = K$$

or $$P = K/V^\gamma$$

$$W = K \int_{V_1}^{V_2} \frac{dV}{V^\gamma}$$

Fig. 1.6

$$= \frac{1}{1-\gamma}\left[\frac{1}{V_2^{\gamma-1}} - \frac{1}{V_1^{\gamma-1}}\right] \quad ...(i)$$

Since A and B lie on the same adiabatic

$$P_1V_1^\gamma = P_2V_2^\gamma = K$$

$$W = \frac{1}{1-\gamma}\left[\frac{1}{V_2^{\gamma-1}} - \frac{1}{V_1^{\gamma-1}}\right]$$

$$W = \frac{1}{1-\gamma}\left[\frac{P_2V_2^{\gamma}}{V_2^{\gamma-1}} - \frac{P_1V_1^{\gamma}}{V_1^{\gamma-1}}\right]$$

$$= \frac{1}{1-\gamma}\left[P_2V_2 - P_1V_1\right] \quad \text{...(ii)}$$

Taking T_1 and T_2 as the temperatures at the points A and B respectively and considering one gram molecule of the gas

$$P_1V_1 = RT_1$$

and $$P_2V_2 = RT_2$$

Substituting these values in equation (ii)

$$W = \frac{1}{1-\gamma}\left[RT_2 - RT_1\right] \quad \text{...(iii)}$$

Here, heat transferred is zero because the system is thermally insulated from the surroundings. The decrease in the internal energy of the system (due to fall in temperature) is equal to the work done by the system and *vice versa*.

1.19 RELATION BETWEEN ADIABATIC AND ISOTHERMAL ELASTICITIES

1. Isothermal Elasticity

During an isothermal process

$$PV = \text{const}$$

Differentiating,

$$PdV + VdP = 0$$

or $$\frac{V.dP}{-dV} = P \quad \text{...(i)}$$

From the definition of elasticity of a gas

$$E_{ise} = \frac{dP}{-dV/V}$$

$$= \frac{VdP}{-dV} \quad \text{...(ii)}$$

From (i) and (ii)

$$E_{ise} = P$$

2. Adiabatic Elasticity

During an adiabatic process

$$PV^{\psi} = \text{const}$$

Differentiating, $P\gamma V^{\gamma-1}dV + V^{\gamma}\,dP = 0$

or $$\frac{VdP}{-dV} = \gamma P \qquad \text{...(iv)}$$

From the definition of elasticity of a gas

$$E_{adt} = \frac{dP}{-dV/V}$$

$$= \frac{V\,.\,dP}{-dV} \qquad \text{...(v)}$$

From (iv) and (v),

$$E_{adt} = \gamma P$$

Comparing (iii) and (vi)

$$E_{adt} = \gamma E_{iso}$$

Thus, the adiabatic elasticity of a gas is γ times the isothermal elasticity.

1.20 CLEMENT AND DESORMES METHOD—DETERMINATION OF γ

Clement and Desormes in 1819 designed an experiment to find γ, the ratio between the two specific heats of a gas.

The vessel A has a capacity of 20 to 30 litres and is fitted in a box containing cotton and wool. At the top end, three tubes are fitted as shown in Fig. 1.7. Through S_1, dry air is forced into the vessel A. The strip cock S_1 is closed when the pressure inside A is slightly greater than the atmospheric pressure. Let the difference in level on the two sides of the manometer be B and the atmospheric pressure be P_0. The pressure of air inside the vend is P_1.

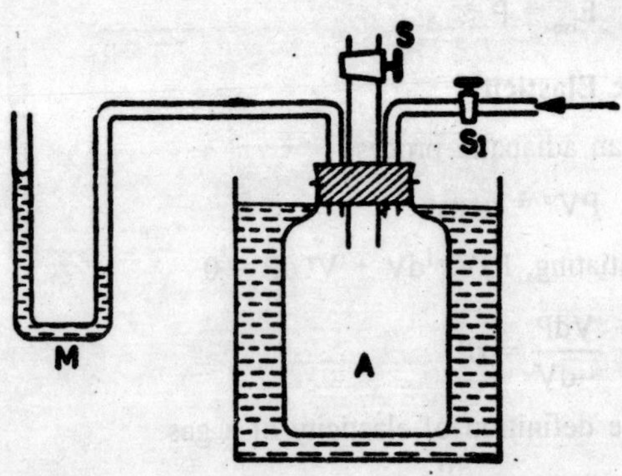

Fig. 1.7

The stop-cock S is suddenly opened and closed just at the moment when the levels of the liquid on the two sides of the manometer are the same. Some quantity of air escapes to the atmosphere. The air inside the vessel expands adiabatically. The temperature of air inside the vessel falls due to adiabatic expansion. The air inside the vessel is allowed to gain heat from the surroundings and it finally attains the temperature of the surroundings. Let the pressure at the end be P_1 and the difference in levels on the two sides of the manometer be h.

Theory: Consider a fixed mass of air left in the vessel in the end. This mass of air has expanded from volume V_1 (less than the volume of the vessel) at pressure P_1 to volume V_2 at pressure P_0. The process is adiabatic as shown by the curve AS (Fig. 6.8).

$$P_1 V_1^{\gamma} = P_0 V_2^{\gamma}$$

$$\frac{P_1}{P_0} = \left(\frac{V_2}{V_1}\right)^{\gamma} \qquad \text{...(i)}$$

Finally the point C a reached. The points A and C are at the room temperature. Therefore AC can be considered as an isothermal.

$$P_1 V_1 = P_2 V_2$$

$$\frac{V_2}{V_1} = \frac{P_1}{P_2} \qquad \text{...(ii)}\backslash$$

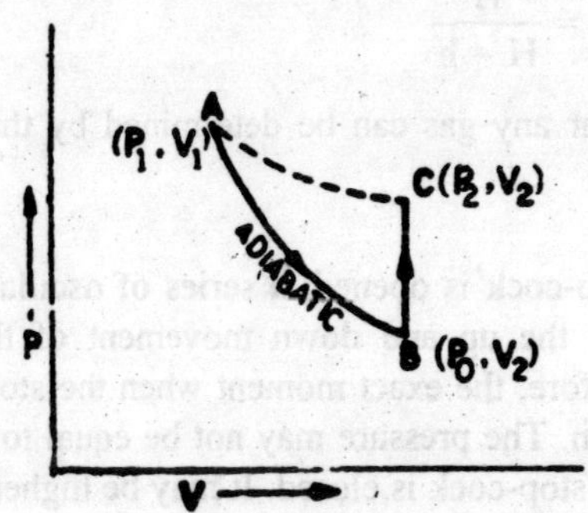

Fig. 1.8

Substituting the value of V_2/V_1 in equation (i),

$$\frac{P_1}{P_0} = \left(\frac{P_1}{P_2}\right)^{\gamma}$$

Taking logarithms,

$$\log P_1 - \log P_0 + \gamma[\log P_1 - \log P_2]$$

$$\gamma = \frac{\log P_1 - \log P_0}{\log P_1 - \log P_2} \qquad \text{...(iii)}$$

But $\qquad P_1 = P_0 + H$ and $P_2 = P_0 + h$

$$\therefore \qquad \gamma = \frac{\log (P_0 - H) - \log P_0}{\log (P_0 - H) - \log (P_0 + h)}$$

$$\gamma = \frac{\log\left(\dfrac{P_0 + H}{P_0}\right)}{\log\left(\dfrac{P_0 + H}{P_0 + h}\right)}$$

$$= \frac{\log\left(1 + \dfrac{H}{P_0}\right)}{\log\left(1 + \dfrac{H - h}{P_0 + h}\right)}$$

Approximately, $\gamma = \dfrac{\dfrac{H}{P_0}}{\dfrac{H - h}{P_0}} = \dfrac{H}{H - h}$

Hence $$\gamma = \frac{H}{H - h}$$ ...(iv)

Similarly, γ far any gas can be determined by this method.

Drawbacks

When the stop-cock is opened, a series of oscillations are set up. This is shown by the up and down movement of the liquid in the manometer. Therefore, the exact moment when the stopcock should be closed h not known. The pressure may not be equal to the atmospheric pressure when the stop-cock is closed. It may be higher or less than the atmospheric pressure. Thus, the result obtained will not be accurate.

1.21 PARTINGTON'S METHOD

Lummer, Pringsheim and Partington designed an apparatus to determine the value of γ. In this method, the pressure and temperature are measured accurately before and after the adiabatic expansion.

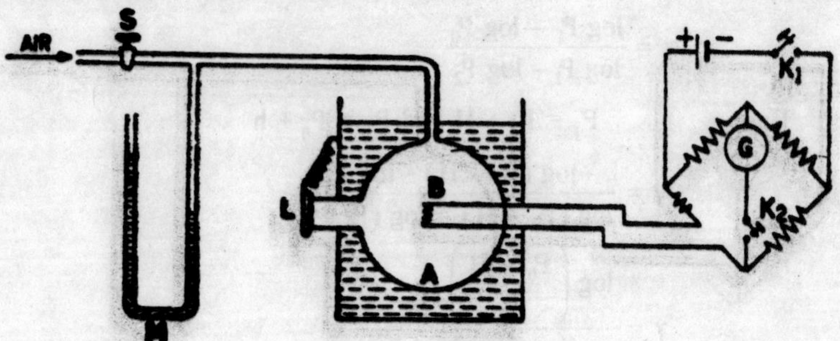

Fig. 1.9

The apparatus consists of a vessel A having a capacity between 130 and 150 litres. The value l can be opened and closed suddenly. It is controlled by a spring arrangement (Fig. 1.10). Dry air (or gas) at a pressure higher than the atmospheric pressure is forced into (he vessel A and the stop-cock S is closed. The oil manometer M is used to measure the pressure of air inside the vessel A. The bolometer B (a platinum wire) and a sensitive galvanometer are used in the Wheatstone's bridge arrangement.

The vessel is surrounded by a constant temperature bath. Let the initial pressure and temperature be P_1 and T_1 (room temperature). The bridge is kept slightly disturbed from the balanced position. The valve L is suddenly opened and closed. The Wheatstones bridge is at once adjusted for balanced position. The temperature of air inside A has decreased due to adiabatic expansion of air. Let the temperature inside be T_0 and the atmospheric pressure P_0. If the apparatus is allowed to remain as such for some time, it will gain heat from the surroundings and the balance point gets disturbed. In order that the balance point remains undisturbed, some pieces of ice are added into the water surrounding the vessel A. When the temperature of water bath is the same as that of air just after adiabatic expansion, the bridge will remain balanced.

The temperature T_0 of the bath represents the temperature of air after the adiabatic expansion.

$$\frac{P_1^{\gamma-1}}{T_1^{\gamma}} = \frac{P_0^{\gamma-1}}{T_0^{\gamma}}$$

$$\left(\frac{P_1}{P_0}\right)^{\gamma-1} = \left(\frac{T_1}{T_0}\right)^{\gamma}$$

$$(\gamma - 1)(\log P_1 - \log P_0) = \gamma\, [\log T_1 - \log T_0]$$

$$\gamma = \frac{\log P_1 - \log P_0}{(\log P_1 - \log P_0) - (\log T_1 - \log T_0)}$$

As P_1, P_0, T_1 and T_0 are known, γ can be calculated. The value of γ for air at 17°C is found to be 1.4034.

Advantages

1. Due to the large volume of the vessel, the expansion is adiabatic.
2. The temperatures are measured accurately just before and after the adiabatic expansion.

Drawbacks

This method cannot be used to find the value of γ at higher temperatures because it is not possible to determine the cooling correction accurately.

1.22 RUCHHARDT'S EXPERIMENT

In 1929, Ruchhardt designed an apparatus to find the value of γ. It is based on the principle of mechanics. Air (or gas) is enclosed in a big jar (Fig. 1.10). A tube of uniform area of cross section is fitted and a ball of mass m fits in the tube just like a piston. In the equilibrium position, the ball is at the point A. The pressure P of air inside the vessel, is given by

$$P = P_0 + mg/A$$

where P_0 is the atmospheric pressure and A is the area of cross-section on of the tube.

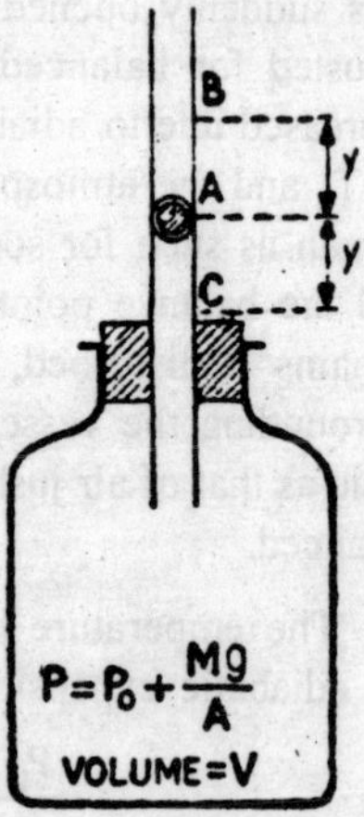

Fig. 1.10

The ball is given a small downward movement through a distance y. It moves up and down and executes simple harmonic motion. Let the change in volume be dV and change in pressure be dP.

$$dV = yA$$

and $$dP = \frac{F}{A}$$

where f is the restoring force

As the ball moves quite rapidly, the process of change in P and V is adiabatic.

$$\therefore \quad PV^{\gamma} = \text{const}$$

Differentiating, $P\gamma V^{\gamma-1}\, dV + V^{\gamma}\,.\, dP = 0$

$$\gamma PdV + VdP = 0$$

Substituting the values of dV and dP,

$$\gamma PyA + \frac{V.F}{A} = 0$$

But $$F = M\,.\,\frac{d^2y}{dt^2}$$

or $$F = -\frac{\gamma PA^2}{V}.y \qquad \text{...(i)}$$

$$\therefore \quad \frac{d^2y}{dt^2} + \frac{\gamma PA^2}{VM}\,.\,y = 0 \qquad \text{...(ii)}$$

Equation (ii) represents S.H.M.

Let the time period be t

$$t = 2\pi \sqrt{\frac{\text{displacement}}{\text{acceleration}}}$$

$$= 2\pi \sqrt{\frac{y}{-d^2y/dt^2}}$$

$$t = 2\pi \sqrt{\frac{VM}{\gamma PA^2}}$$

$$\gamma = 2\pi \frac{4\pi^2 \ VM}{PA^2t^2} \quad \text{...(iii)}$$

As F, M, P and A are already known, only the time period t is to be determined.

Hence γ can be calculated.

This experiment is not very accurate because the ball comes to rest in a short time due to friction and the time period cannot be determined accurately.

Rinkle's Modification

Rinkle suggested a modification in Ruchhardt's experiment. The ball is kept in the equilibrium position where the pressure of air (or gas) is exactly equal to the atmospheric pressure. The ball is allowed to drop and the distance L through which it drops before starting to move up again is noted. This distance is found accurately by taking a series of photographs of the moving ball.

$$\text{Work done} = \int_0^L F.dy$$

$$\text{But} \quad F = \frac{\gamma P_0 A^2 y}{V}$$

$$W = \frac{\gamma P_0 A^2}{V} \int_0^L ydy$$

$$= \frac{\gamma P_0 A^2 L^2}{2V} \quad \text{...(iv)}$$

$$\text{Work done} = MgL \quad \text{...(v)}$$

Equating (iv) and (v)

$$\frac{\gamma P_0 A^2 L^2}{2V} = MgL$$

$$\gamma = \frac{2MgV}{P_0 A^2 L} \qquad ...(vi)$$

As M, g, V, P_0 and A are known, L is measured from the photographic film. Hence y can be calculated. The error in the result is due to the following assumptions : (i) that there is no friction, (ii) that the gas is ideal and (iii) the process is completely adiabatic.

1.23 IRREVERSIBLE PROCESS

The thermodynamical state of a system can be defined with the help of the thermodynamical coordinates of the system. The state of a system can be changed by altering the thermodynamical coordinates. Changing from one state to the other by changing the thermodynamical coordinates is called a process.

Consider two states of a system *i.e.*, state A and state B. Change of state from A to B or vice versa is a process and the direction of the process will depend upon a new thermodynamical coordinate called entropy. All processes are not possible in the universe..

Consider the following processes :

1. Let two blocks A and B at different temperatures T_1 and T_2 ($T_1 > T_2$) be kept in contact but the system as a whole is insulated from the surroundings. Conduction of heat takes place between the blocks, the temperature of A falls and the temperature of B rises and thermodynamical equilibrium will be reached.
2. Consider a flywheel rotating with an angular velocity ω. Its initial kinetic energy is $1/2I\omega^2$. After some time the wheel comes to rest and kinetic energy is utilised in overcoming friction at the bearings. The temperature of the wheel and the bearings rises and the increase in their internal energy is equal to the original kinetic energy of the fly wheel.
3. Consider two flasks A and B connected by a glass tube provided with a stop cock. Let A contain air at high pressure and B is evacuated. The system is isolated from the surroundings. If the stop cock is opened, air rushes from A to B, the pressure in A decreases and the volume of air increases.

All the above three examples though different, are thermodynamical processes involving change in thermodynamical coordinates. Also, in accordance with the first law of thermodynamics, the principle of conservation of energy is not violated because the total energy of the system is conserved. It is also clear that, with the initial conditions described above, the three processes will take place.

Let us consider the possibility of the above three processes taking place in the reverse direction. In the first case, if the reverse process is possible, the block B should transfer heat to A and initial conditions should be restored. In the second case, if the reverse process is possible, the heat energy must again change to kinetic energy and the fly wheel should start rotating with the initial angular velocity ω. In the third case, if the reverse process is possible the air in B must flow back to A and the initial condition should be obtained.

But, it is a matter of common experience, that none of the above conditions for the reverse processes are reached. It means that the direction of the process cannot be determined by knowing the thermodynamical coordinates in the two end states. To determine the direction of the process a new thermodynamical coordinate has been devised by Clausius and this is called the entropy of the system. Similar to internal energy, entropy is also a function of the state of a system. For any possible process, the entropy of an isolated system should increase or remain constant. The process in which there is a possibility of decrease in entropy cannot take place.

If the entropy of an isolated system is maximum, any change of state will mean decrease in entropy and hence that change of state will not take place.

To conclude, *processes in which the entropy of an isolated system decreases do not take place or for all processes taking place in an isolated system the entropy of the system should increase or remain constant*. It means a process is irreversible if the entropy decreases when the direction of the process is reversed. A process is said to be irreversible if it cannot be retraced back exactly in the opposite direction. During an irreversible process, heat energy is always used to overcome friction. Energy is also dissipated in the form of conduction and radiation. This loss of energy always takes place whether the engine works in one direction or the reverse direction. Such energy cannot be regained. In actual practice all the engines are irreversible. If electric current is passed through a wire, heat is produced. If the direction of the current is reversed, heat is again produced. This is also an example of an irreversible process. All chemical reactions are irreversible. In general, all natural processes are irreversible.

1.24 REVERSIBLE PROCESS

From the thermodynamical point of view, a reversible process is one in which an infinitesimally small change in the external conditions will result in all the changes taking place in the direct process but exactly repeated in the reverse order and in the opposite sense. The process should take place at an extremely slow rate. In a reversible cycle, there should not be any loss of heat due to friction or radiation. In this process, the initial conditions of the working substance can be obtained.

Consider a cylinder, containing a gas at a certain, pressure and temperature. The cylinder is fitted with a frictionless piston. If the pressure is decreased, the gas expands slowly and maintains a constant temperature (isothermal process). The energy required for this expansion is continuously drawn from the source (surroundings). If the pressure on the piston is increased, the gas contracts slowly and maintains constant temperature (isothermal process). The energy liberated during compression is given to the sink (surroundings). This is also true for an adiabatic process provided the process takes place infinitely slowly.

The process will not be reversible if there is any loss of heat due to friction, radiation or conduction. If the changes take place rapidly, the process will not be reversible. The energy used in overcoming friction cannot be retraced.

The conditions of reversibility for any heat engine or process can be stated as follows:

1. The pressure and temperature of the working substance must not differ appreciably from those of the surroundings at any stage of the cycle of operation.
2. All the processes taking place in the cycle of operation must be infinitely slow.
3. The working parts of the engine must be completely free from friction.
4. There should not be any loss of energy due to conduction or radiation during the cycle of operation.

It should be remembered that the complete reversible process or cycle of operation is only an ideal case. In an actual process, there is always loss of heat due to friction, conduction or radiation. The temperature and pressure of the working substance differ appreciably from those of the surroundings.

1.25 SECOND LAW OF THERMODYNAMICS

A heat engine is chiefly concerned with the conversion of heat energy into mechanical work. A refrigerator is a device to cool a certain space below the temperature of its surroundings. The first law of thermodynamics is a qualitative statement which does not preclude the possibility of the existence of either a heat engine or a refrigerator. The first law does not contradict the existence of a 100% efficient heat engine or a self-acting refrigerator.

In practice, these two are not attainable. These phenomena are recognized and this led to the formulation of a law governing these two devices. It is called second law of thermodynamics.

A new term reservoir is used to explain the second law. A *reservoir* is a device having infinite thermal capacity and which can absorb, retain or reject unlimited quantity of heat without any change in its temperature.

Kelvin-Planck statement of the second law is as follows:

"It is impossible to get a continuous supply of work from a body (or engine) which can transfer heat with a single heat reservoir. This is a negative statement. According to this statement, a single reservoir at a single temperature cannot continuously transfer heat into work. It means that there should be two reservoirs for any heat engine. One reservoir (called the source) is taken at a higher temperature and the other reservoir (called the sink) is taken at a lower temperature.

According to this statement, zero degree absolute temperature is not attainable because no heat is rejected to the sink at zero degree Kelvin. If an engine works between any temperature higher than zero degree Kelvin and zero degree Kelvin, it means it uses a single reservoir which contradicts Kelvin-Planck's statement of the second law. Similarly, no engine can be 100% efficient.

In a heat engine, the engine draws heat from the source and after doing some external work, it rejects the remaining heat to the sink. The source and sink are of infinite thermal capacity and they maintain constant temperature.

First Part

According to Kelvin, the second law can also be stated as follows :

"It is impossible to get a continuous supply of work from a body by cooling it to a temperature lower than that of its surroundings".

In a heat engine the working substance does some work and rejects the remaining heat to the *sink*. The temperature of the source must be higher than the surroundings and the engine will not work when the temperatures of the source and the sink are the same. Take the case of a steam engine. The steam (working substance) at high pressure is introduced into the cylinder of the engine. Steam expands, and it does external work. The contents remaining behind after doing work are rejected to the surroundings. The temperature of the working substance rejected to the surroundings is higher than the temperature of the surroundings.

If this working substance rejected by the first engine is used in another engine, it can do work and the temperature of the working substance will fall further.

It means that the working substance can do work only if its temperature is higher than that of the surroundings.

Second Part

According to Clausius:

"It is impossible to make heat flow from a body at a lower temperature to a body at a higher temperature without doing external work on the working substance."

This part is applicable in the case of ice plants and refrigerators. Heat itself cannot flow from a body at a lower temperature to a body at a higher temperature. But, it is possible, if some external work is done on the working substance. Take the case of ammonia ice plant. Ammonia is the working substance. Liquid ammonia at low pressure takes heat from the brine solution in the brine tank and is converted to low pressure vapour. External work is done to compress the ammonia vapours to high pressure. This ammonia at high pressure is passed through coils over which water at room temperature is poured. Ammonia vapour gives heat to water at room temperature and gets itself converted into liquid again. This high pressure liquid ammonia is throttled to low pressure liquid ammonia. In the whole process ammonia (the working substance) takes heat from brine solution (at a lower temperature) and gives heat to water at room temperature (at a higher temperature). This is possible only due to the external work done on ammonia by the piston in compressing it. The only work of electricity in the ammonia ice plant is to move the piston to do external work on ammonia. If the external work is not done, no ice plant or refrigerator will work. Hence, it is possible to make heat flow from a body at a lower temperature to a body at a higher temperature by doing external work on the working substance.

Thus, the second law of thermodynamics plays an important part for practical devices *e.g.*, heat engines and refrigerators. The first law of thermodynamics only gives the relation between the work done and the heat produced. But the second law of thermodynamics gives the conditions under which heat can be converted into work.

1.26 CARNOT'S REVERSIBLE ENGINE

Heat engines are used to convert heat into mechanical work. Sadi Carnot (French) conceived a theoretical engine which is free from all the defects of practical engines. Its efficiency is maximum and it is an ideal heat engine.

For any engine, there are three essential requisites:

1. **Source:** The source should be at a fixed high temperature T_1 from which the heat engine can draw heat. It has infinite thermal capacity and any amount of heat can be drawn from it at constant temperature T_1.
2. **Sink:** The sink should be at a fixed lower temperature T_2 to which any amount of heat can be rejected. It also has infinite thermal capacity and its temperature remains constant at T_2,
3. **Working Substance:** A cylinder with non-conducting sides and conducting bottom contains the perfect gas as the working substance.

A perfect non-conducting and frictionless piston is fitted into the cylinder. The working substance undergoes a complete cyclic operation (Fig. 1.11).

A perfectly non-conducting stand is also provided so that the working substance can undergo adiabatic operation.

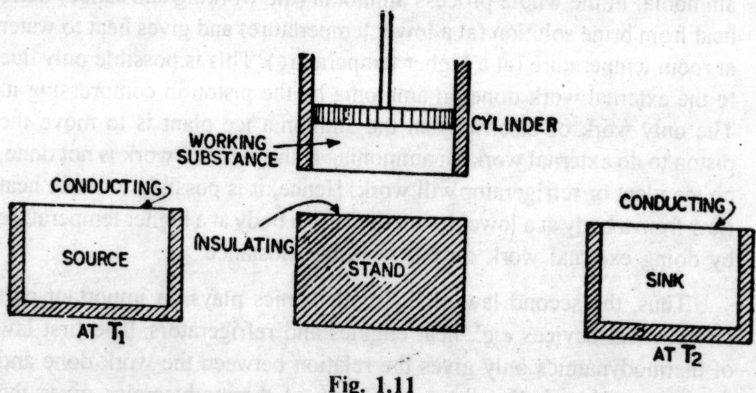

Fig. 1.11

Carnot's Cycle

1. Place the engine containing the working substance over the source at temperature T_1. The working substance is also at a temperature T_1. Its pressure is Pi and volume is V_1 as shown by the point A in Fig. 1.12. Decrease the pressure. The volume of the working substance increases. Work is done by the working substance. As the bottom is perfectly conducting to the source at temperature T_1, it absorbs heat. The process is completely isothermal. The temperature remains constant. Let the amount of heat absorbed by the working substance be H_1 at the temperature T_1. The point B is obtained.

   Consider one gram molecule of the working substance.

   Work done from A to B (isothermal process)

   $$W_1 = \int_{V_1}^{V_2} P.dV = RT_1 \log \frac{V_2}{V_1}$$

   $$= \text{area ABGE.} \qquad \text{...(i)}$$

2. Place the engine on the stand having an insulated top. Decrease the pressure on the working substance. The volume increases. The process is completely adiabatic. Work is done by the working substance at the cost of its internal energy. The temperature falls. The working substance undergoes adiabatic change from B to C. At C the temperature is T_2 (Fig. 1.12).

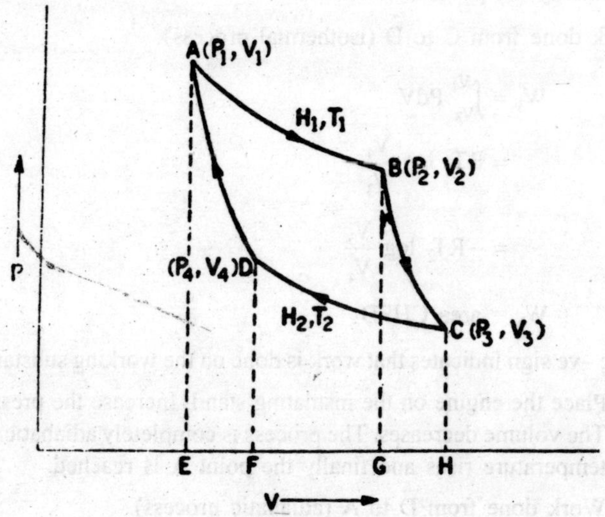

Fig. 1.12

Work done from B to C (adiabatic process)

$$W_2 = \int_{V_2}^{V_3} P.dV$$

$$= \int_{V_2}^{V_3} \frac{dV}{V^\gamma}$$

$$= \frac{KV_3^{1-\gamma} - KV_2^{1-\gamma}}{1-\gamma}$$

But PV^γ = constant = K

$P_2V_2 = RT_1$

$P_3V_3 = RT_2$

$P_3V_3^\gamma = P_2V_2^\gamma = K$

$$= \frac{P_3V_3 - P_2V_2}{1-\gamma}$$

$$= \frac{R[T_2 - T_1]}{1-\gamma} = \frac{R[T_1 - T_2]}{\gamma - 1}$$

W_2 = Area BCHG ...(ii)

3. Place the engine on the sink at temperature T_2. Increase the pressure. The work is done on the working substance. As the base is conducting to the sink, the process is isothermal. A

quantity of heat H_2 is rejected to the sink at temperature T_1. Finally the point D is reached.

Work done from C to D (isothermal process)

$$W_3 = \int_{V_3}^{V_4} PdV$$

$$= RT_2 \log \frac{V_4}{V_3}$$

$$= -RT_2 \log \frac{V_3}{V_4}$$

$$W_3 = \text{area CHFD}$$

(The –ve sign indicates that work is done on the working substance.)

4. Place the engine on the insulating stand. Increase the pressure. The volume decreases. The process is completely adiabatic. The temperature rises and finally the point A is reached.

   Work done from D to A (adiabatic process).

$$W_4 = \int_{V_4}^{V_1} PdV$$

$$= -\frac{R(T_1 - T_2)}{\gamma - 1}$$

$$W_4 = \text{area DFEA} \quad ...(iv)$$

[W_2 and W_4 are equal and opposite and cancel each other.]

The net work done by the working substance in one complete cycle

= Area ABGE + Area BCHG – Area CHFD – Area DFEA

= Area ABCD

The net amount of heat absorbed by the working substance

$$= H_1 - H_2$$

$$\text{Network} = W_1 + W_2 + W_3 + W_4$$

$$= RT_1 \log \frac{V_2}{V_1} + \frac{R(T_1 T_2)}{\gamma - 1} - RT_2 \log \frac{V_3}{V_4} - \frac{R[T_1 - T_2]}{\gamma - 1}$$

$$W = RT_1 \log \frac{V_2}{V_1} - RT_2 \log \frac{V_3}{V_4} \quad ...(v)$$

The points A and D are on the same adiabatic

$$T_1 V_1^{\gamma-1} = T_2 V_4^{\gamma-1}$$

$$\frac{T_2}{T_1} = \left(\frac{V_1}{V_4}\right)^{\gamma-1} \quad \text{...(vi)}$$

The points B and C are on the same adiabatic

$$T_1 V_2^{\gamma-1} = T_2 V_3^{\gamma-1}$$

$$\frac{T_2}{T_1} = \left(\frac{V_2}{V_3}\right)^{\gamma-1} \quad \text{...(vii)}$$

From (vi) and (vii)

$$\left(\frac{V_1}{V_4}\right)^{\gamma-1} = \left(\frac{V_2}{V_3}\right)^{\gamma-1}$$

or $$\frac{V_1}{V_4} = \frac{V_2}{V_3}$$

or $$\frac{V_2}{V_1} = \frac{V_3}{V_4}$$

From equation (v)

$$W = RT_1 \log \frac{V_2}{V_1} - RT_2 \log \frac{V_2}{V_1}$$

$$W = R\left[\log \frac{V_2}{V_1}\right][T_1 - T_2]$$

$\therefore$ $$W = H_1 - H_2$$

Efficiency $$\eta = \frac{\text{Useful output}}{\text{Input}} = \frac{W}{H_1}$$

Heat is supplied from the source from A to B only.

$$H_1 = RT_1 \log \frac{V_2}{V_1}$$

$\therefore$ $$\eta = \frac{W}{H_1} = \frac{H_1 - H_2}{H_1}$$

$$= \frac{R[T_1 - T_2] \log \left(\frac{V_2}{V_1}\right)}{RT_1 \log \left(\frac{V_2}{V_1}\right)}$$

or $$\eta = 1 - \frac{H_2}{H_1}$$

$$\eta = 1 - \frac{T_2}{T_1} \qquad \text{...(viii)}$$

The Carnot's engine is perfectly reversible. It can be operated in the reverse direction also. Then it works as a refrigerator. The heat H_2 is taken from the sink and external work is done on the working substance and heat H_1 is given to the source at a higher temperature.

The isothermal process will take place only when the piston moves very slowly to give enough time for the heat transfer to take place. The adiabatic process will take place when the piston moves extremely fast to avoid heat transfer satisfy these conditions.

All practical engines have an efficiency less than the Carnot's engine.

1.27 CARNOT'S ENGINE AND REFRIGERATOR

Carnot's cycle is perfectly reversible. It can work as a heat engine and also as a refrigerator. When it works as a heat engine, it absorbs a quantity of heat H_1 from the source at a temperature T_1, does an amount of work W and rejects an amount of heat H_2 to the sink at temperature T_2. When it works as a refrigerator, it absorbs heat H_2 from the sink at temperature T_2. W amount of work is done on it by some external means and rejects heat H_1 to the source at a temperature T_1 (Fig. 1.13). In the second case heat flows from a body at a lower temperature to a body at a higher temperature, with the help of external work done on the working substance and it works as a refrigerator. This will not be possible if the cycle is not completely reversible.

Coefficient of Performance

The amount of heat absorbed at the lower temperature is H_2. The amount of work done by the external process (input energy) = W and the amount of heat rejected = H_1. Here H_2 is the desired refrigerating effect.

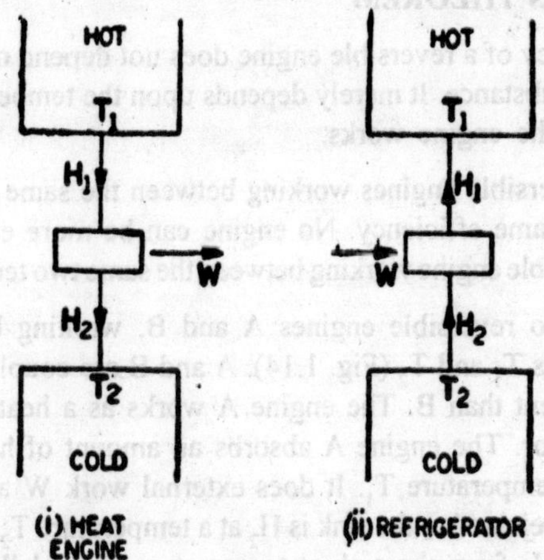

Fig. 1.13

Coefficient of performance

$$= \frac{H2}{W} = \frac{H_2}{H_1 - H_2}$$

Suppose 200 joules of energy is absorbed at the lower temperature and 100 joules of work is done with external help. Then 200 + 100 = 300 joules are rejected at the higher temperature.

The coefficient of performance

$$= \frac{H2}{W}$$

$$= \frac{H_2}{H_1 - H_2}$$

$$= \frac{200}{300 - 200} = 2.$$

Therefore the coefficient of performance of a refrigerator = 2.

In the case of a heat engine, the efficiency cannot be more than 100% but in the case of a refrigerator, the coefficient of performance can be much higher than 100%.

1.28 CARNOT'S THEOREM

The efficiency of a reversible engine does not depend on the nature of the working substance. It merely depends upon the temperature limits between which the engine works.

"All the reversible engines working between the same temperature limits have the same efficiency. No engine can be more efficient than a Carnot's reversible engine working between the same two temperatures."

Consider two reversible engines A and B, working between the temperature limits T_1 and T_2 (Fig. 1.14). A and B are coupled. Suppose A is more efficient than B. The engine A works as a heat engine and B as a refrigerator. The engine A absorbs an amount of heat H_1 from the source at a temperature T_1. It does external work W and transfers it to B. The heat rejected to the sink is H_2 at a temperature T_2. The engine B absorbs heat H_1 from the sink at temperature T_2 and W amount of work is done on the working substance. The heat given to the source at temperature T_1 is H_1'.

Suppose the engine A w more efficient than B.

Efficiency of the engine A

$$= \eta = \frac{H_1 - H_2}{H_1} = \frac{W}{H_1}$$

Efficiency of the engine B

$$= \eta' = \frac{H_1' - H_2'}{H_1'} = \frac{W}{H_1'}$$

Since $\eta' > \eta'$; $H_1' > H_2'$

Also $W = H_1 - H_2 = H_1' - H_2'$

$\therefore \quad H_2' > H_2$

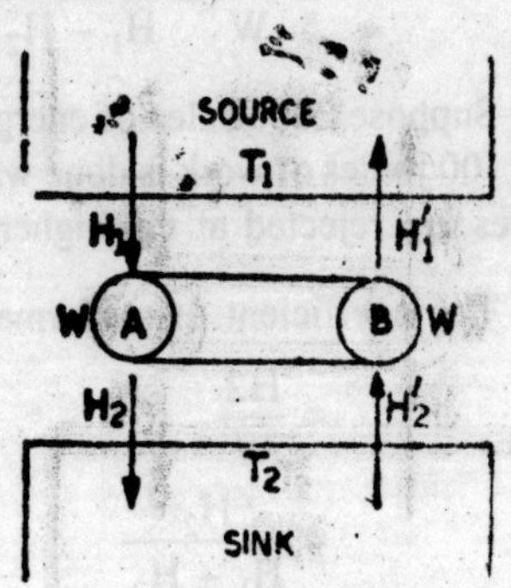

Fig. 1.14

Thus, for the two engines A and B working as a coupled system, $(H_2' - H_2)$ is the quantity of heat taken from the sink at a temperature T_2 and $(H_1' - H_1)$ is the quantity of heat given to the source at a temperature T_1. Both $(H_2' - H_2)$ and $(H_1' - H_1)$ are positive quantities. It means heat flows from the sink at a temperature T_2 (lower temperature) to the source at a temperature T_1 (higher temperature) *i.e.*, heat flows from a body at a lower temperature to a body at a higher temperature. But, no external work has been done on the system. This is contrary to the second law of thermodynamics. Thus, η cannot be greater than η'. The two engines (reversible) working between the same two temperature limits have the same efficiency. Moreover, in the case of a Carnot's engine, there is no loss of heat due to friction, conduction or radiation (irreversible processes). Thus, the Carnot's engine has the maximum efficiency. Whatever may be the nature of the working substance, the efficiency depends only upon the two temperature limits.

In a practical engine there is always loss of energy due to friction, conduction, radiation etc. and hence its efficiency is always lower than that of a Carnot's engine.

1.29 THERMODYNAMIC (OR WORK OR ABSOLUTE) SCALE OF TEMPERATURE

The efficiency of a Carnot's engine does not depend upon the property of the working substance but it depends upon the temperature limits between which the engine works. The scale of temperature based on the working of the Carnot's engine is a standard scale and it does not depend upon the particular property of any substance, as in the case of other thermometric scales. Kelvin worked out the theory of the absolute scale called Kelvin's or thermodynamical scale and also showed that it agrees with the perfect gas scale.

Suppose an engine works between the temperatures θ_1 and θ_2. Here H_1 is the heat absorbed at θ_1 and H_2 is the heat rejected at θ_2

$$\eta = f(\theta_1, \theta_2)$$

$$h = \frac{H_1 - H_2}{H_1} = f(\theta_1, \theta_2)$$

$$\therefore \quad 1 - \frac{H_1}{H_2} = f(\theta_1, \theta_2)$$

or
$$\frac{H_1}{H_2} = \frac{1}{1 - f(\theta_1, \theta_2)} = F(\theta_1, \theta_2) \quad \text{...(i)}$$

Similarly, if the engine works between the temperature θ_2 and θ_3. and H_2 is the heat absorbed and H_3 is the heat rejected, then

$$\frac{H_2}{H_3} = F(\theta_2, \theta_2) \quad \text{...(ii)}$$

If the engine works between the temperatures θ_1 and θ_3 and H_1 is the heat absorbed and H_3 is the heat rejected, then

$$\frac{H_1}{H_3} = F(\theta_1, \theta_3) \qquad \text{...(iii)}$$

From (i), (ii) and (iii),

$$\frac{H_1}{H_3} = \frac{H_1}{H_2} \times \frac{H_2}{H_3}$$

$$\therefore \qquad F(\theta_1, \theta_3), F(\theta_1, \theta_2) \times F(\theta_2, \theta_3)$$

$$F(\theta_1, \theta_2) = \frac{F(\theta_1, \theta_3)}{F(\theta_2, \theta_3)}$$

This temperature θ_3 is arbitrarily chosen.

This relation can be satisfied only if

$$F(\theta_1, \theta_2) = \frac{\phi(\theta_1)}{\phi(\theta_2)}$$

$$F(\theta_2, \theta_3) = \frac{\phi(\theta_2)}{\phi(\theta_3)}$$

and $$F(\theta_1, \theta_3) = \frac{\phi(\theta_1)}{\phi(\theta_3)}$$

where ϕ is another function.

$$\therefore \qquad \frac{H_1}{H_2} = F(\theta_1, \theta_2) = \frac{\phi(\theta_1)}{\phi(\theta_2)}$$

This temperature θ_3 is arbitrarily chosen. This relation can be satisfied only if

$$\therefore \qquad \frac{H_1}{H_2} = \frac{\theta_1}{\theta_2} \qquad \text{...(iv)}$$

The expression on the right hand side represents the ratio of the two Kelvin temperatures and this can be denoted as $\frac{\theta_1}{\theta_2}$.

The relation (iv) is used to represent a new scale and this does not depend upon the property of the working substance.

$$\text{Efficiency} \qquad \eta = \frac{H_1 - H_2}{H_1} = \frac{\theta_1 - \theta_2}{\theta_1} \qquad \text{...(v)}$$

$$\eta = 1 - \frac{\theta_2}{\theta_1} \quad \text{....(vi)}$$

1.30 ABSOLUTE ZERO ON WORK SCALE

Consider an engine working between the steam point and the ice point. The isothermal AB is at the steam point and the isothermal CD is at the ice point (Fig. 1.15).

$$\eta = 1 - \frac{\theta_{ice}}{\theta_{steam}}$$

$$\eta = \frac{\theta_{steam} - \theta_{ice}}{\theta_{steam}}$$

If the distance between the isothermals AB and CD is divided into 100 equal parts by drawing isothermals parallel to AB or CD, then any isothermal will be at a temperature 1° higher than the isothermal just below it. The difference in temperature between the steam point and the ice point is 100 on this scale also. In this way, the work scale (Kelvin) of temperature is fixed. Similarly, isothermals below the ice point can be drawn and an isothermal representing absolute zero can be obtained (Fig. 1.15). The isothermal GK represents absolute zero.

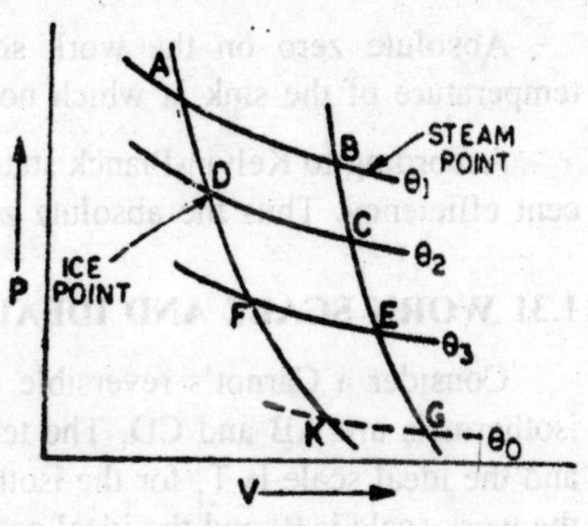

Fig. 1.15

Consider an engine working between the steam point and absolute zero ($\theta_0 = 0$).

$$\eta = 1 - \frac{\theta_0}{\theta_{steam}} = 1$$

i.e., the engine has efficiency equal to 1.

Similarly an engine working between ice point and absolute zero, will have an efficiency.

$$\eta' = 1 - \frac{\theta_0}{\theta_{ice}} = 1$$

But efficiency $\eta = \dfrac{H_1 - H_2}{H_1} = 1 - \dfrac{H_2}{H_1}$

If $\quad 1 - \dfrac{H_2}{H_1} = 1, \ \dfrac{H_2}{H_1} = 0$

or $\quad H_2 = 0$

It means that if the sink is at the temperature of zero degree absolute, no heat is rejected and the whole of the available energy has been used up in doing useful work and in this case the engine will have 100% efficiency.

Thus the absolute zero temperature on the work scale is defined as the temperature at which a system undergoes a reversible isothermal process without any transfer of heat.

Absolute zero on the work scale can also be defined as that temperature of the sink at which no heat is rejected to it.

According to Kelvin-Planck statement, no engine can have 100 per cent efficiency. Thus the absolute zero temperature is not attainable.

1.31 WORK SCALE AND IDEAL GAS SCALE

Consider a Carnot's reversible cycle ABCD (Fig. 1.16). The two isothermals are AB and CD. The temperature on the work scale is θ_1 and the ideal scale is T_1 for the isothermal AB and the temperature on the work scale is θ_2 and the ideal gas scale is T_2 for the isothermal CD. Let the pressure and volume for the points A, B, C and D be P_1, V_1; P_2, V_2; P_3, V_3 and P_4, V_4 respectively. Consider that 1 gram of an ideal gas is used as the working substance.

From A to B, the process is isothermal.

$$\text{Work done} = \int_{V_1}^{V_2} P dV$$

$$= rT_1 \int_{V_1}^{V_2} \frac{dV}{V}$$

$$\therefore \quad H_1 = rT_1 \log \frac{V_2}{V_1} \quad \text{...(i)}$$

Here r is expressed in heat units.

From C to D, the process is isothermal and a quantity of heat H_2 is rejected by the gas.

$$H_2 = -\int_{V_3}^{V_4} PdV$$

$$= -rT_2 \int_{V_3}^{V_4} \frac{dV}{V}$$

$$H_2 = rT_2 \log \frac{V_3}{V_4} \quad \text{...(ii)}$$

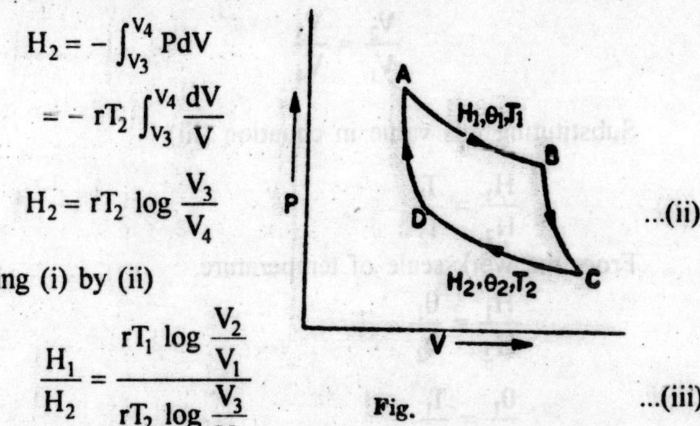

Fig.

Dividing (i) by (ii)

$$\frac{H_1}{H_2} = \frac{rT_1 \log \frac{V_2}{V_1}}{rT_2 \log \frac{V_3}{V_4}} \quad \text{...(iii)}$$

Also, A and B lie on the same isothermal

$$P_1V_1 = P_2V_2$$

or $$\frac{P_2}{P_1} = \frac{V_1}{V_2} \quad \text{...(iv)}$$

C and D lie on the same isothermal

$$P_3V_3 = P_4V_4$$

or $$\frac{P_3}{P_4} = \frac{V_4}{V_3} \quad \text{...(v)}$$

B and 0 lie on the same adiabatic

$$P_2V_2^{\gamma} = P_3V_3^{\gamma} \quad \text{...(vi)}$$

A and D lie on the same adiabatic

$$P_1V_1^{\gamma} = P_4V_4^{\gamma} \quad \text{...(vi)}$$

Dividing (vi) by (vii)

$$\left(\frac{P_2}{P_1}\right)\left(\frac{V_2}{V_1}\right)^{\gamma} = \left(\frac{P_3}{P_4}\right)\left(\frac{V_3}{V_4}\right)^{\gamma}$$

or $$\left(\frac{V_1}{V_2}\right)\left(\frac{V_2}{V_1}\right)^{\gamma} = \left(\frac{V_4}{V_3}\right)\left(\frac{V_3}{V_4}\right)^{\gamma}$$

$$\left(\frac{V_2}{V_1}\right)^{\gamma-1} = \left(\frac{V_3}{V_4}\right)^{\gamma-1}$$

or

$$\frac{V_2}{V_1} = \frac{V_3}{V_4} \qquad ...(viii)$$

Substituting this value in equation (iii)

$$\frac{H_1}{H_3} = \frac{T_1}{T_2} \qquad ...(ix)$$

From the work scale of temperature,

$$\frac{H_1}{H_2} = \frac{\theta_1}{\theta_2} \qquad ...(x)$$

$$\therefore \quad \frac{\theta_1}{\theta_2} = \frac{T_1}{T_2} \qquad ...(xi)$$

Special considerations

(i) If θ_1 is zero, T_1 is also zero. It means the two scales are identical at absolute zero temperature.

(ii) Consider that the Carnot's engine works between the steam point and the ice point and there are 100 degrees between the ice point and the steam point on the two scales.

$$\frac{\theta_{steam}}{\theta_{ice}} = \frac{T_{steam}}{T_{ice}}$$

$$\frac{\theta_{ice} + 100}{\theta_{ice}} = \frac{T_{ece} + 100}{T_{ice}}$$

or $\qquad \theta_{ice} = T_{ice}$

Therefore, the ice point Is the same on the two scales.

Similarly $\theta_{steam} = T_{steam}$

Hence the two scales are completely identical.

Efficiency: The efficiency of a Carnot's engine,

$$\eta = 1 - \frac{H_2}{H_1}$$

$$= 1 - \frac{\theta_2}{\theta_1} = 1 - \frac{T_2}{T_1}$$

Let the adiabatic expansion ratio be ρ

$$\rho = \frac{V_3}{V_2}$$

The points B and C lie on the same adiabatic

$$\therefore \quad T_1 V_2^{\gamma-1} = T_2 V_3^{\gamma-1}$$

$$\frac{T_2}{T_1} = \left(\frac{V_2}{V_3}\right)^{\gamma-1}$$

$$= \left(\frac{1}{\rho}\right)^{\gamma-1}$$

$$\eta = 1 - \frac{T_2}{T_1} = 1 - \left(\frac{1}{\rho}\right)^{\gamma-1}$$

1.32 RANKINE CYCLE

In a steam engine, water is used as the working substance. Water is converted into steam in the boiler. The steam is super-heated above the boiling temperature. This superheated steam is introduced into the cylinder. For the first part, steam from the boiler enters the cylinder and moves the piston. The inlet valve remains open and the steam enters at a constant pressure. In the second part, the inlet valve closes and the steam expands adiabatically. The piston is moved through the rest of the working stroke. As soon as the piston reaches the end of *the stroke* the exhaust valve opens and the steam goes to the condenser. The steam gets condensed in the condenser.

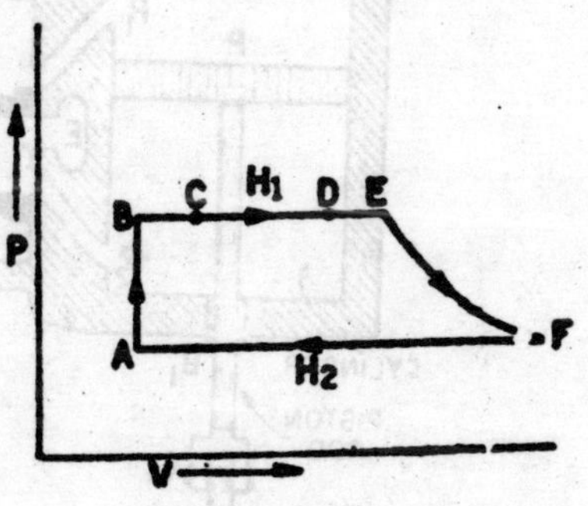

Fig. 1.17

This cycle of operations is known as the Rankine cycle, which is an ideal case. Its working is divided into the following parts:

1. A to B (Fig. 1.17) represents the compression of water and the boiler pressure is reached.
2. B to C represents the heating of water in the boiler at constant pressure and the boiling point is reached.

3. C to D represents conversion of water into steam at constant pressure.
4. D to E represents superheating of steam beyond its boiling point.
5. E to f represents the adiabatic expansion of steam in doing external work. Steam is cooled and gets condensed.
6. F to A represents the rejection of steam to the condenser where the whole of the steam gets condensed. Finally, the point A is reached.

Efficiency: From B to E, the amount of heat absorbed is H_1 and from F to A, the amount of heat rejected is H_2.

$$\eta = 1 - \frac{H_2}{H_1}$$

1.33 STEAM ENGINE

It consists of a steam chest and a cylinder. These are connected by the steam ports P_1 and P_2. P is a piston that moves inside the cylinder.

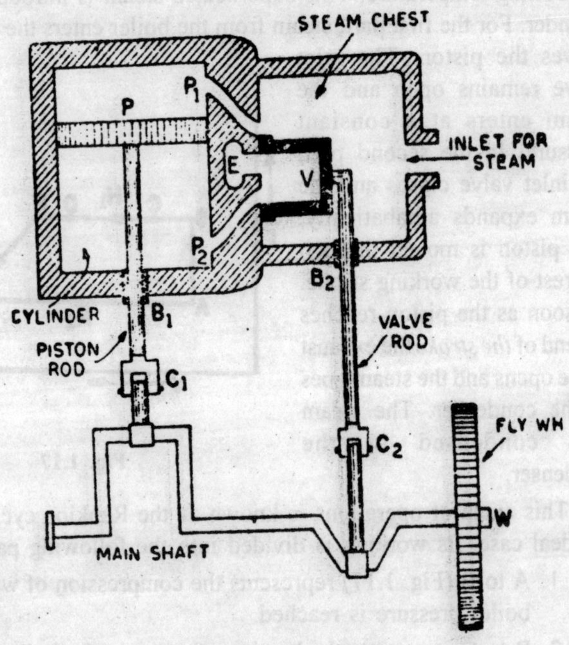

Fig. 1.18

V is a sliding valve which can also move to and fro such that the port P_1 or P_2 is open to the steam chest. The piston and the sliding valve are connected to the main shaft eccentrically at diametrically opposite points such that when the shaft rotates they move in opposite directions. W is a flywheel attached to the main shaft. The coupling between the main rod and the piston roe is such that the linear motion of the piston rod is converted to circular motion of the shaft. The machinery to be operated is connected to the main shaft.

Working

1. Steam at high pressure from the boiler enters the steam chest. The steam enters the cylinder through the port P_1. The piston P is pushed down due to the high pressure of steam (Fig. 1.18). The sliding valve moves up. Both the ports P_1 and P_2 are closed. The steam inside the cylinder expands and moves the piston further in the downward direction. At this position, the port P_1 is connected to the exhaust valve E and P_2 is opened.
2. When steam enters through the port P_3, the piston is pushed up and the sliding valve moves down.

The dead steam in the cylinder is exhausted out to the atmosphere through the exhaust valve E. Both the ports P_1 and P_2 are closed and the steam in the cylinder expands. The piston is pushed further to complete the stroke. Finally the port P_2 is closed and P_1 opens. The process is continuously repeated.

B_1 and B_2 are the stuffing boxes. These boxes allow for the free and smooth sliding of the valve rod and the piston rod. C_1 and C_2 are the cross heads. When the piston moves to and fro, the shaft is rotated and finally the machinery rotates.

Dead Centres

Twice in the rotation of the shaft, the force acts at right angles to the shaft. Theoretically a force does not produce any effect when it acts at right angles to the shaft. For this purpose *the flywheel* is used so that the momentum gaïned by the flywheel enables the shaft to rotate beyond the dead centre in each half rotation. Thus, the flywheel makes the motion of the shaft smooth and not jerky.

Power

Suppose the average pressure of steam = P newtons/metre^2. The length of the stroke = L metres and the area of cross section of the piston = A sq metres.

Total force on the piston = (PA) newtons.

Distance moved by the piston in a double acting steam engine = 2L

$\therefore$ Work done for one rotation of the shaft = (PA) × 2 L joules.

Let the shaft rotate N times per second.

Work done per second = (PA) × 2L × N watts

$$\therefore \text{Power of the engine} = \frac{2PLAN}{1000} \text{ Kilowatts}$$

Efficiency

Most of the heat energy in the steam engine is wasted. Therefore, its efficiency is very low. It is of the order of 8 to 10%. Efficiency is the ratio of the useful mechanical work (W) done by the engine to the amount of heat (H) supplied by the fuel for the same time.

1.34 OTTO CYCLE

In the Otto engine (internal combustion engine), air is the working substance and petrol vapour acts as the fuel. A set of ideal conditions are assumed while discussing an Otto cycle:

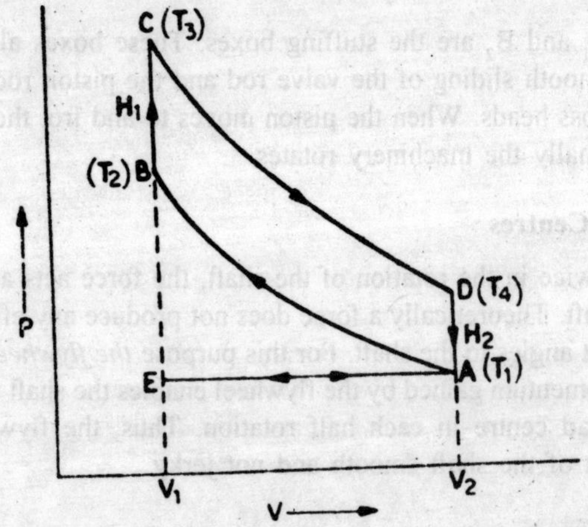

Fig. 1.19

1. The working substance is air all the time and it behaves as a perfect gas.

2. There is no friction.

On the basis of these assumptions, Otto cycle can be divided into the following operations:

1. E to A (Fig. 1.19) represents the *charging stroke*. Thc mixture of air and petrol vapour is allowed to enter at atmospheric pressure. The final volume is V_2 at temperature T_1.

2. A to S represents the *adiabatic compression*. There is no friction and no flow of heat through the walls of the cylinder. The volume changes from V_2 to V_1 and the temperature changes from T_1 to T_2. There is change in pressure also. This represents the *compression, stroke*. The compression ratio is nearly 8 *i.e.*, $V_2/V_1 = 8$ nearly. The pressure changes from one atmosphere to about 18 atmospheres.
3. B to G represents the *ignition stage*. A spark is produced and the mixture of air and petrol vapour is ignited. The pressure increases from 18 to about 80 atmospheres and the temperature changes from T_2 to T_3 (from 500°C to 2000°C).
4. C to D represents the *working stroke*. The gas expands adiabatically and the engine works. The volume changes from V_1 to V_2. The pressure and temperature decrease.
5. As the point D is reached, the exhaust valve opens. D to A represents the change of pressure to the atmospheric pressure and the temperature changes from T_4 to T_1.
6. A to E represents the *exhaust stroke*. The exhaust gases are completely discharged from the cylinder. Thus, the initial condition of the engine is restored.

Efficiency : Consider that 1 gram of the working substance is used in the process. A quantity of heat H_1 at a higher temperature is absorbed from B to C,

$$H_1 = 1 \times C_v (T_3 - T_2).$$

A quantity of heat Hy is rejected at^ lower temperature from D to A

$$H_2 = 1 \times C_v (T_4 - T_1).$$

$$\frac{H_2}{H_1} = \frac{T_4 - T_1}{T_3 - T_2} \quad ...(i)$$

$$\eta = 1 - \frac{H_2}{H_1}$$

$$= 1 - \frac{T_4 - T_1}{T_3 - T_2} \quad ...(ii)$$

The points D and C lie on the same adiabatic

$$\therefore \quad T_4 V_2^{\gamma-1} = T_3 V_1^{\gamma-1} \quad ...(iii)$$

The points A and B lie on the same adiabatic

$$T_1 V_2^{\gamma-1} = T_2 V_1^{\gamma-1} \quad ...(iv)$$

Subtracting (iv) from (iii),

$$(T_4 - T_1) V_2^{\gamma-1} = (T_3 - T_2) V_1^{\gamma-1}$$

$$\frac{T_4 - T_1}{T_2 - T_2} = \left(\frac{V_1}{V_2}\right)^{\gamma-1} \quad ...(v)$$

Substituting this value in equation (ii),

$$\eta = 1 - \left(\frac{V_1}{V_2}\right)^{\gamma-1}$$

$$= 1 - \left(\frac{1}{(V_2/V_1)}\right)^{\gamma-1}$$

$$\eta = 1 - \left(\frac{1}{\rho}\right)^{\gamma-1} \quad ...(vi)$$

where ρ is the adiabatic compression ratio.

In an actual petrol engine p cannot be made greater than 10. If ρ is more than 10, the mixture gets ignited by itself due to compression much before the sparking takes place.

$$\eta = 1 - \left(\frac{1}{\rho}\right)^{\gamma-1}$$

If $\quad \rho = 6$ and $\gamma = 1.4$

$$\eta = 1 - \left(\frac{1}{6}\right)^{0.4}$$

$$= 0.5116 = 51.16\%$$

If $\rho = 9$ and $\gamma = 1.4$

$$\eta = 1 - \left(\frac{1}{9}\right)^{0.4}$$

$$= 0.5847 = 58.47\%.$$

1.35 INTERNAL COMBUSTION ENGINE (PETROL ENGINE)

It consists of a cylinder with an air-tight piston. It has two valves. I is the inlet valve and O is the outlet valve. The valves are opened and closed at the proper times with a suitable mechanism. The petrol vapours and air from the *carburettor* are admitted into the cylinder through the valve I. Here S acts as a sparking plug. It is also known as a four-stroke engine.

1. Charging Stroke

The mixture of petrol vapours and air enters the cylinder. The piston moves down [Fig. 1.20 (i)].

2. Compression Stroke

The inlet valve closes and the piston moves up. The mixture is compressed to a high pressure and the temperature rises to about 500°C [Fig. 1.20 (ii)].

3. Working Stroke

A spark is produced at the sparking plug and the compressed mixture of petrol and air ignites. The temperature rises to about 2000°C and a high pressure of about 15 atmospheres is developed inside the cylinder. The piston is pushed down and the shaft rotates [Fig. 1.20 (iii)]. Only in this stroke, work is done by the piston.

4. Exhaust Stroke

The outlet valve opens and the burnt and unburnt gases are exhausted out to the atmosphere [Fig. 1.20 (iv)].

The cycle repeats and the engine works.

Engine Cooling

The internal combustion engine requires cooling, although the efficiency of the engine will be higher without cooling. Temperatures of the order of 2000°C are reached inside the cylinder, although this temperature may last only for a very short interval of time before expansion. If the engine is not cooled the following drawbacks will be present:

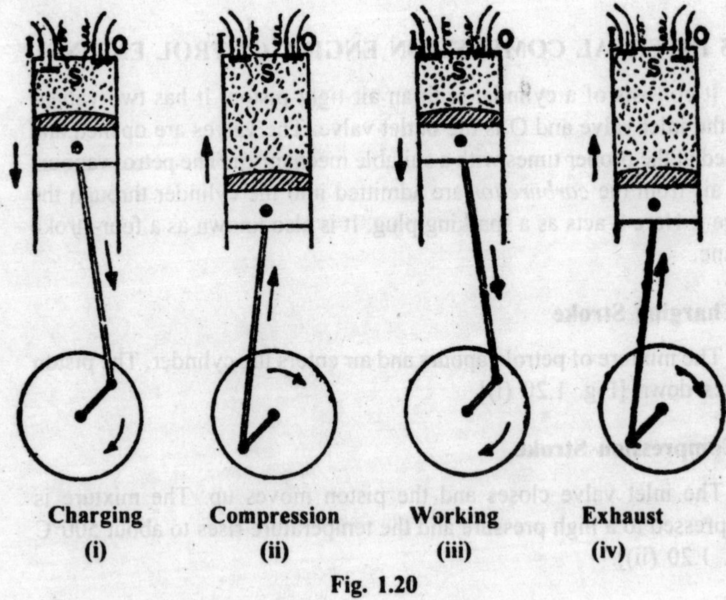

Fig. 1.20

(1) Burning of the piston and exhaust valves.

(2) Irregular burning of the fuel.

(3) Seizure of the piston due to expansion.

(4) Burning of the lubricant, and

(5) Melting of the bearing material.

1.36 DIESEL ENGINE

In the case of a diesel engine, air is admitted into the cylinder in the beginning. This air is compressed adiabatically so that the temperature is high enough to ignite the oil sprayed into the cylinder. The ideal conditions are assumed while discussing the complete cycle *viz.* (i) the working substance is air all the time and it behaves as a perfect gas and (ii) there is no friction. The cycle is called the *diesel cycle.*

1. E to A (Fig. 1.21) represents the intake of air and its volume changes to V_1 This is the *charging stroke.*
2. A to B represents the *compression stroke.* Air is compressed adiabatically. The temperature changes from T_1 to T_2 and the volume changes from V_1 to V_2. The temperature riles to about 1000°C and the pressure to about 40 atmospheres.

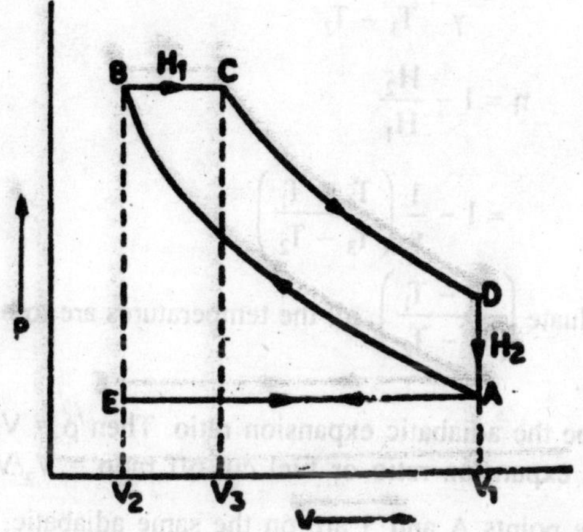

Fig. 1.21

3. S to 0 represents the stage when the oil is sprayed into the cylinder. The oil burns immediately. The pressure is maintained constant. The temperature changes from T_2 to T_3 (to 2000°C) and the volume changes from V_2 to V_3.
4. C to D represents the *working stroke.* The mixture of air and diesel oil vapour expand adiabatically.
5. As the point D is reached, the exhaust valve opens and the pressure drops to the point A. The volume remains constant, but temperature and pressure decrease.
6. A to E represents the *exhaust stroke.* The unburnt vapours of the oil and the mixture of gases in the cylinder are exhausted out of the cylinder.

Efficiency: From B to C, the pressure remains constant. Considering 1 gram of the working substance, the quantity of heat absorbed,

$$H_1 = 1 \times C_P (T_3 - T_2)$$

From D to A the volume remains constant.

The quantity of heat rejected,

$$H_2 = 1 \times C_V (T_4 - T_1)$$

$$\frac{H_2}{H_1} = \frac{C_V}{C_P} \cdot \frac{(T_4 - T_1)}{(T_3 - T_2)}$$

$$= \frac{1}{\gamma} \cdot \frac{T_4 - T_1}{T_3 - T_2}$$

$$\eta = 1 - \frac{H_2}{H_1}$$

$$= 1 - \frac{1}{\gamma}\left(\frac{T_4 - T_1}{T_3 - T_2}\right) \quad ...(1)$$

To evaluate $\left(\frac{T_4 - T_1}{T_3 - T_2}\right)$, all the temperatures are to be expressed in terms of T_2.

Let ρ be the adiabatic expansion ratio. Then $\rho = V_1/V_2$ and e the combustion expansion ratio or fuel cut off ratio $= V_3/V_2$.

(1) The points A and 5 are on the same adiabatic,

$$\therefore \quad T_1 V_1^{\gamma-1} = T_2 V_2^{\gamma-1}$$

$$T_1 = T_2 \left[\frac{V_2}{V_1}\right]^{\gamma-1}$$

$$= T_2 \left[\frac{1}{\rho}\right]^{\gamma-1} \quad ...(2)$$

(2) The points B and C are at the same pressure

$$\therefore \quad \frac{T_3}{V_3} = \frac{T_2}{V_2}$$

or $$T_3 = T_2\left[\frac{V_3}{V_2}\right] = T_2\,[e] \qquad ...(3)$$

(3) The points C and D are on the same adiabatic.

$$\therefore \quad T_4V_4^{\gamma-1} = T_3V_3^{\gamma-1} \qquad [\text{But } V_4 = V_1]$$

$$T_4 = T_3\left[\frac{V_3}{V_1}\right]^{\gamma-1}$$

$$= T_3\left[\frac{V_3}{V_2}\times\frac{V_2}{V_1}\right]^{\gamma-1}$$

$$T_4 = T_3\left[\frac{V_3}{V_2}\right]^{\gamma-1}\left[\frac{V_2}{V_1}\right]^{\gamma-1}$$

$$T_4 = T_3\,[e]^{\gamma-1}\left[\frac{1}{\rho}\right]^{\gamma-1}$$

$$T_4 = T_2\,[e]\,[e]^{\gamma-1}\left[\frac{1}{\rho}\right]^{\gamma-1}$$

$$T_4 = T_2\,[e]^{\gamma}\left[\frac{1}{\rho}\right]^{\gamma-1} \qquad ...(4)$$

$$\eta = 1 - \left[\frac{T_4 - T_1}{T_3 - T_2}\right]$$

$$= 1 - \frac{1}{\gamma}\left[\frac{T_2\,[e]^{\gamma}\left[\frac{1}{p}\right]^{\gamma-1} - T_2\left[\frac{1}{p}\right]^{\gamma-1}}{T_3\,[e] - T_2}\right]$$

$$\eta = 1 - \frac{1}{\gamma}\left(\frac{1}{\rho}\right)^{\gamma-1}\left[\frac{e^{\gamma-1}}{e-1}\right] \qquad ...(5)$$

For the same compression ratio, the efficiency of an Otto engine is more than a diesel engine. In practice, the compression ratio for an

Otto engine is from 7 to 9 and for a diesel engine it is from 15 to 20. Due tot he higher compression ratio, an actual diesel engine has higher efficiency than the Otto (Petrol) engine. The cylinder must be strong enough to withstand very high pressure.

1.37 MULTICYLINDER ENGINES

With an engine having one cylinder, the engine works only during the working stroke. The piston moves during the rest of the three strokes due to the momentum of the shaft. In a multicylinder engine (say 4-cylinder engine) the four cylinders are coupled. The working of each cylinder is given below:

	First	Second	Third	Fourth
First quarter	Working	Exhaust	Compression	Charging
Second quarter	Exhaust	Charging	Working	Compression
Third quarter	Charging	Compression	Exhaust	Working
Fourth quarter	Compression	Working	Charging	Exhaust

In this way, the power of the engine increases and the shaft gets momentum during each quarter cycle.

1.38 CLAPEYRON LATENT HEAT EQUATION

Consider the isothermals FBAE at temperature T + dT and GCDH at temperature T. Here EA and HD show the liquid state of the substance. At A to B or D to C the substance is in transition from the liquid to the gaseous state and *vice versa.* At B and C the substance is purely in the gaseous state. From B to F or C to G the substance is in the gaseous state. Join A to D and B to C by dotted lines.

The cycle ABCD represents a complete cycle and Carnot's theorem can be applied. Suppose the volume at the point A is V_1 and temperature is T = dT. The pressure is just below its saturation pressure and the liquid begins to evaporate and at the point B the volume is V_2. The substance is in the vapour state. Suppose the mass of the liquid at B is one gram. The amount of heat absorbed is H_1, Here H_1 = L + dL, where L + dL is the latent heat of the liquid at temperature (T + dT).

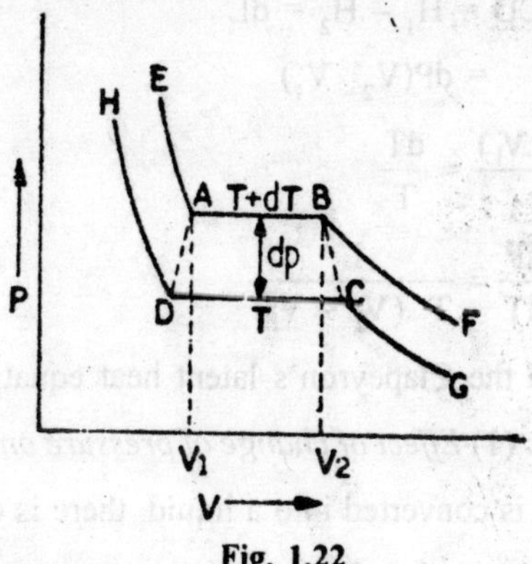

Fig. 1.22

At the point B, the pressure is decreased by dP. The vapour will expand and its temperature falls. The temperature at C is T. As this pressure and temperature T, the gas begins to condense and is converted into the liquid state. At the point D, the substance is in the liquid state. From C to D, the amount of heat rejected (given out) is H_2 Here H_2 = L where L is the latent heat at temperature T. By increasing the pressure a little, the original point A is restored. The cycle ABCDA is completely reversible. Applying the principle of the Carnot's reversible cycle

$$\frac{H_1}{T_1} = \frac{H_2}{T_2}$$

or $$\frac{H_1}{H_2} = \frac{T_1}{T_2}$$

$$\frac{H_1 - H_2}{H_2} = \frac{T_1 - T_2}{T_2}$$

Here, $H_1 = L + dL,, H_2 = L$

$T_1 = T + dT,, T_2 = T$

$H_1 - H_2 = L + dL - L = dL$

$T_1 - T_2 = T + dT - T = dT$

$$\therefore \quad \frac{dL}{L} = \frac{dT}{T}$$

The area of the figure

$$ABCD = H_1 - H_2 = dL$$

$$= dP(V_2 - V_1)$$

$$\therefore \quad \frac{dP\,(V_2 - V_1)}{L} = \frac{dT}{T}$$

$$\frac{dP}{dT} = \frac{L}{T\;(V_2 - V_1)} \qquad \text{...(i)}$$

This is called the Clapeyron's latent heat equation.

Applications: (1) *Effect of change of pressure on the melting point.*

When a solid is converted into a liquid, there is change in volume.

(i) If V_2 is greater than V_1

dP/dT is a positive quantity. It means that the rate of change of pressure with respect to temperature is positive. In such cases, the melting point of the substance will increase with increase in pressure and vice versa.

(ii) If V_2 is less than V_1.

dP/dT is a negative quantity. It means that the rate of change of pressure with respect to temperature is negative. In such cases, the melting point of the substance will decrease with increase in pressure and *vice versa.* In the case of melting ice, the volume of water formed is less than the volume of ice taken. Hence $V_2 < V_1$.

Therefore, the melting point of ice decreases with increase in pressure. 'Hence ice will melt at a temperature lower than zero degree centigrade at a pressure higher than the normal pressure.

Ice melts at 0°C only at a pressure of 76 cm of Hg.

(2) *Effect of change of pressure on the boiling point.*

When a liquid is converted into a gaseous state, the volume V_2 of the gas is always greater than the corresponding volume V_1 of the liquid *i.e.* $V_2 > V_1$.

Therefore, dP/dT is a +ve quantity.

With increase in pressure, the boiling point of a substance increases and *vice versa.* The liquid will boil at a lower temperature under reduced pressure. In the case of water, the boiling point increases with increase in pressure and vice- versa. Water boils at 100°C only at 76 cm of Hg pressure. In the laboratories, while preparing steam, the boiling point is less than 100°C because the atmospheric pressure is less than 76 cm of Hg. In pressure cookers, the liquid boils at a higher temperature because the pressure inside is more than the atmospheric pressure.

1.39 THERMIONIC EMISSION

The electrical conductivity of metals is due to the free electrons present in the metal and not attached to any particular molecule. These free electrons continuously move within the metal and the velocity of the electrons increases with the increase in temperature of the metal. These electrons exert a pressure similar to that of a gas. However, the electrons cannot escape to the outer space because of the attractive forces at the surface that tend to keep the electrons within the metal. The attractive forces are much greater than the *electron gas pressure*. In order to escape from the surface, an electron has to do a certain amount of work called the work function (ϕ) of the surface and this is different for different metals. This energy is obtained from the kinetic energy of the electrons. At Ordinary temperatures the kinetic energy of the electrons is much less than the work function and hence the electrons cannot escape from the surface. With the increase in temperature of the metal surface, the kinetic energy of the electrons increases and if the kinetic energy exceeds the work function, electrons can escape from the surface of the metal. This phenomenon is called *thermionic* emission and the electrons are called *thermo-electrons* or *thermions*. The thermionic current (I) is given by the equation

$$I = AT^{e}e^{-\frac{\phi_0}{kT}}$$

where A is a constant, T is the absolute temperature of the metal, ϕ_0 is the work function of the surface at zero degree K and k is the Boltzmann's constant. I measures the thermionic current in amperes per sq cm of the emitting surface. The most commonly used emitters are *(i) tungsten (ii) thoriated tungsten and (iii) oxide coated cathodes containing barium or strontium.*

1.40 RICHARDSON'S EQUATION

Richardson derived the equation for the thermionic emission from a heated metal. He applied Clausius-Clapeyron's latent heat equation for the ejection of electrons from the surface of the metal.

Let the volume for 1 gram molecule of electrons within the metal be v_1 and outside the metal be v_2. The volume v_2 is very high as compared to v_1. The pressure of electrons under equilibrium conditions is P. The energy absorbed by N electrons during ejection from the metal surface = W. Here N is the Avogadro's number. Applying Clapeyron's latent heat equation

$$\frac{dP}{dT} = \frac{W}{T(v_2 - v_1)} \quad ...(i)$$

As $v_2 >> v_1$, v_1 can be neglected

$$\therefore \quad \frac{dP}{dT} = \frac{W}{T\,v_1}$$

or $$W = T\,v_2 \frac{dP}{dT} \quad ...(ii)$$

The energy W consists of two parts :

(i) The work function of the surface *i.e.*, energy required for ejecting N electrons just outside the metal *i.e.*, in crossing the potential barrier. This energy = $N\phi$ where ϕ is the work function of the surface.

(ii) The work done in admitting the ejected electrons into the electron cloud already present outside the metal. This energy is Pv_2. For one gram molecule of electrons

$$Pv_2 + RT = N\,k\,T$$

Here k is Boltzmann's constant

$$W = N\,\phi + N\,k\,T \quad ...(iii)$$

From equations (ii) and (iii)

$$N\,(\phi + kT) = Tv_2 \frac{dP}{dT}$$

But $$v_2 = \frac{NkT}{P}$$

$$\therefore N\,(\phi + kT) = T\left(\frac{NkT}{P}\right) \times \frac{dP}{dT}$$

$$(\phi + kT) = \frac{kT^2}{P} \times \frac{dP}{dT}$$

$$\frac{dP}{P} = \left(\frac{\phi + kT}{kT^2}\right) dT$$

Integrating

$$\int \frac{dP}{P} = \int \frac{\phi}{kT^2}\, dT + \int \frac{dT}{T}$$

$$\log_e P = \int \frac{\phi}{kT^2}\, dT + \log e\ T + \text{constant}$$

Let the constant be equal to $\log_e A_1$

$$\log_e P - \log_e T - \log_e A_1 = \int \frac{\phi}{kT^2}\, dT$$

$$\log_e \left(\frac{P}{A_1 T}\right) = \int \frac{\phi}{kT^2}\, dT \qquad \text{...(iv)}$$

Richardson suggested that ϕ also depends upon T. If ϕ_0 is the work function at 0 K, then

$$\phi = \phi_0 + 3/2kT$$

$$\therefore \int \frac{\phi}{kT^2}\, dT = \int \frac{\phi_0 + 3/2kT}{kT^2}$$

$$= \frac{\phi_0}{kT} + \frac{3}{2} \log_e T$$

$$= \frac{\phi_0}{kT} + \log_e T^{3/2}$$

Substituting this value in equation (iv)

$$\log_e \left(\frac{P}{A_1 T}\right) = -\frac{\phi_0}{kT} + \log_e T^{3/2}$$

$$\log_e \left(\frac{P}{A_1 T^{3/2}}\right) = -\frac{\phi_0}{kT}$$

or

$$P = A_1\, T^{3/2}\, e^{-\frac{\phi_0}{kT}} \qquad \text{...(v)}$$

If the number of electrons per unit volume = n then

$$P = n\, k\, T \qquad \text{...(vi)}$$

Equating (v) and (vi)

$$n k T = A_1 T^{3/2} e^{-\frac{\phi_0}{kT}}$$

$$n = \left(\frac{A_1}{k}\right) T^{3/2} e^{-\frac{\phi_0}{kT}} \quad \text{...(vii)}$$

Assuming the perfect gas laws to hold good in the case of an electron cloud, the mean energy of the electron,

$$\frac{1}{2} m_0 v^2 = \frac{3}{2} kT$$

or
$$v = \left(\frac{3kT}{m_0}\right)^{1/2}$$

But Kundsen has shown that the number of electrons passing unit area in one second is,

$$n_0 = \frac{nv}{\sqrt{6\pi}}$$

$$\therefore \quad n_0 = n \times \left(\frac{3kT}{6\pi m_0}\right)^{1/2}$$

$$= n \times \left(\frac{kT}{2\pi m_0}\right)^{1/2}$$

Substituting the value of n from equation (vii)

$$n_0 = \left[\left(\frac{A_1}{k}\right) T^{3/2} e^{-\frac{\phi_0}{kT}}\right]\left[n T^{1/2} \left(\frac{k}{2\pi m_0}\right)^{1/2}\right]$$

$$n_0 = \left[\frac{nA_1}{k} \times \left(\frac{k}{2\pi m_0}\right)^{1/2}\right] T^2 e^{-\frac{\phi_0}{kT}}$$

Taking

$$\frac{nA_1}{k} \times \left(\frac{k}{2\pi m_0}\right)^{1/2} = A_2 = \text{constant}$$

$$n_0 = A_2 T^2 e^{-\frac{\phi_0}{kT}}$$

Taking the charge on the electron = e, the current density due to electron emission,

$$I = n_0 e = A_2 T^2 e^{-\frac{\phi_0}{kT}}$$

$$I = AT^2 e^{-\frac{\phi_0}{kT}} \quad \text{...(viii)}$$

Here $A = (A_2 e)$ = constant.

Equation (viii) represents the Richardson's equation for thermionic emission.

The electron emitter can be heated directly or indirectly. In the case of direct heating, the current is passed through the filament which itself serves as the cathode. Tungsten and thoriated tungsten filaments are the commonly used directly heated electron emitters. The indirectly heated electron emitter consists of a heater wire surrounded by a metal sleeve in the form of a cylinder whose surface is coated with electron emitting materials. The most commonly used cathode is the indirectly heated cathode. In both the cases, filaments can be heated either by DC or AC.

The operating temperatures for tungsten, thoriated tungsten and oxide coated emitters are given below.

Tungsten	2200 to 3000 K
Thoriated Tungsten	1900 K
Oxide coated cathode	1000 to 1150 K.

Values of the Emission Constant

Material	ϕ_0 in eV	A in A/m²–K	Melting point in K
Carbon	4.7	–	–
Nickel	5.0	2.68×10^5	1725
Platinum	5.32	3.20×10^5	2047
Tantalum	4.1	6.02×10^5	3123
Tungsten	4.52	6.02×10^5	3655

1.41 CLAUSIUS INEQUALITY

Before considering entropy, it is necessary to establish the Clausius inequality which states that

$$\oint \frac{\delta H}{T} < 0$$

This relation is between the temperatures of an arbitrary number of heat reservoirs and the amounts of heat transferred (given up or lost).

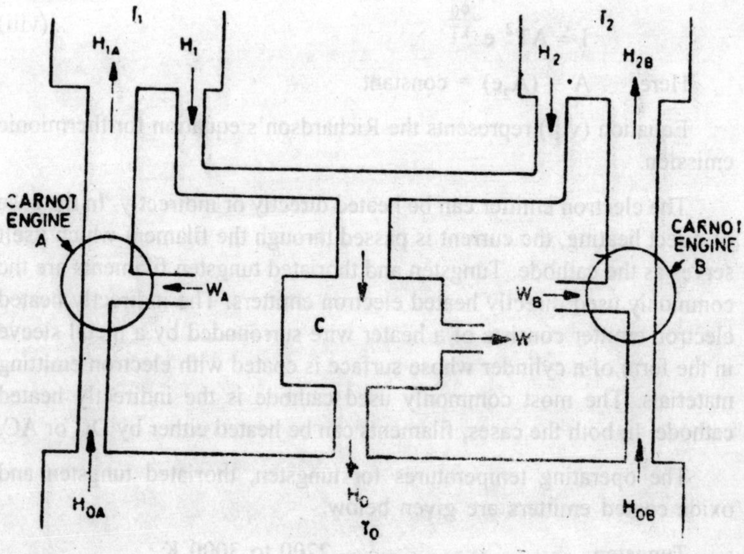

Fig. 1.23

Consider three heat reservoirs at temperatures T_1, T_2 and T_0 (Fig. 1.23).

The engine C works as a heat engine. It absorbs heat H_1 from reservoir at temperature T_1 and H_2 from reservoir at temperature T_2. The system as a whole does an amount of work W and rejects a quantity of heat By to the reservoir at temperature T_0.

Carnot's engines A and B work as refrigerators. The engine A absorbs heat H_{0A} from the reservoir at temperature T_0, an amount of work W_A supplied to it and it rejects a quantity of heat H_{1A} to the reservoir at temperature T_1. The engine B absorbs heat H_{0B} from the reservoir at temperature T_0, an amount of work W_B is supplied to it and it rejects a quantity of heat H_{2B} to the reservoir at temperature T_2.

In the general discussion given below, no restriction on the direction of interchanges of the energy is made except that the interchanges must be consistent with the first and the second laws of thermodynamics.

The engine A working as a refrigerator rejects heat H_{1A} to the reservoir at temperature T_1.

Here, $H_{1A} + H_i = 0$ ...(i)

Similarly, the engine B working as a refrigerator rejects heat H_{2B} to the reservoir, at temperature T_2.

Here $H_{2B} + H_2 = 0$ ...(ii)

It means that when the three engines work simultaneously, there is no outstanding change in the temperature of reservoirs at temperatures T_1 and T_2.

Thus, there has been no change except for the reservoir at temperature T_0 and the mechanical system that supplied or absorbed work. If H_0 is not equal to the sum of H_{0A} and H_{0B}, the reservoir at temperature T_0 must have gained or lost heat. Similarly, if W is not equal to the sum of W_A and W_B, the mechanical system must have gained or lost work. According to the first law, the net flow of heat from or to a heat reservoir must necessarily be equal to the amount of work done on or by the mechanical system. As far as the first law is concerned it is a matter of indifference whether the reservoir gains or loses heat as long as the mechanical system loses or gains the same quantity of work. The consideration is different with reference to the second law.

If the heat reservoir loses heat and the mechanical system gains an equal amount of work, it means that the whole of the heat absorbed from the reservoir is converted into work but this violates the Kelvin-Planck statement of the second law of thermodynamics.

Therefore, H_0 must always be greater than the sum of H_{0A} and H_{0B} except in the limiting case when they are equal.

For the engine A,

$$\frac{H_{1A}}{T_1} = \frac{H_{0A}}{T_0} \quad \text{...(ii)}$$

For the engine B,

$$\frac{H_{2A}}{T_2} = \frac{H_{0B}}{T_0} \quad \text{...(iv)}$$

Also the quantities, H_1 and H_2 are positive, H_0 is negative, H_{1A} and H_{aB} are negative, H_{0A}. and B_{0B} are positive. Mathematically we write all the quantities as positive.

From equation (iii)

$$H_{0A} \frac{T_0(H_{1A})}{T_1} = \frac{T_0(H_1)}{T_1} = T_0\left(\frac{H_1}{T_1}\right) \quad ...(v)$$

Similarly, $$H_{0B} \frac{T_0(H_{2A})}{T_2} = \frac{T_0(H_2)}{T_2} = T_0\left(\frac{H_2}{T_2}\right) \quad ...(vi)$$

The net quantity of heat drawn from the reservoir at temperature T_0 is

$$H_0 + H_{0A} + H_{0B}$$

According to the second law, this quantity should be less than or equal to zero.

i.e. $$H_0 + H_{0A} + H_{0B} < 0$$

Substituting the values of H_{0A} and H_{0B}

$$H_0 + T_0\left(\frac{H_1}{T_1}\right) + T_0\left(\frac{H_2}{T_2}\right) < 0$$

Dividing by T_0

$$\frac{H_0}{T_0} + \frac{H_1}{T_1} + \frac{H_2}{T_2} < 0 \quad ...(vii)$$

In general, for a system using a large number of reservoirs, it can be written as,

$$\sum \frac{H}{T} < 0 \quad ...(viii)$$

When the number of reservoirs is infinite and the heat exchanges take place by infinitesimal amounts, it can be written as

$$\int \frac{\delta H}{T} < 0 \quad ...(ix)$$

Equations (viii) and (ix) represent the *Clausius inequality.* The relation holds goods for any cyclic process, reversible or irreversible.

Examples:

(1) Consider two blocks at temperatures 300 K and 200 K that are in contact. The block at higher temperature conducts heat to the block at-the lower temperature. Suppose, 200 joules of heat is transferred. Here,

$$H_1 = 200 \text{ joules}, \qquad T_1 = 300 \text{ K}$$

H_2 = 200 joules, T_2 = 200 K

$$\therefore \quad \sum \frac{H}{T} = \frac{H_1}{T_1} + \frac{H_2}{T_2}$$

$$= \frac{200}{300} + \frac{(-200)}{200} = -\frac{1}{3}$$

$= -0.33$ joule/degree.

(2) Consider an actual heat engine working between the temperatures 500 K and 300 K.

Suppose its efficiency is 20% and it takes 1000 joules of heat from the high temperature reservoir. As the efficiency is 20%.

$$\eta = 1 - \frac{H_2}{H_1}$$

$$\frac{200}{100} = 1 - \frac{H_2}{100}$$

or H_2 = 800 joules

$$\therefore \quad \sum \frac{H}{T} = \frac{H_1}{T_1} + \frac{H_2}{T_2}$$

Here H_1 = + 1000 joules

H_2 = –800 joules (since heat is rejected)

T_1 = 500 K

T_2 = 300 K

$$\sum \frac{H}{T} = \frac{1000}{500} + \frac{-800}{300}$$

= –2/3 joule/degree.

(3) Consider a Carnots reversible engine working between the temperatures 500 K and 300 K. Suppose 1000 joules of heat energy is drawn from the high temperature reservoir.

Here $$\frac{H_1}{T_1} = \frac{H_2}{T_2}$$

$$\frac{1000}{500} = \frac{H_2}{300}$$

or $\quad H_2 = 600$ joules

$$\sum \frac{H}{T} = \frac{H_1}{T_1} + \frac{H_2}{T_2}$$

Here $\quad H_1 = +\ 1000$ joules

$H_2 = -600$ joules

$T_1 = 500K$

$T_2 = 300K$

$$\therefore \quad \sum \frac{H}{T} = \frac{1000}{500} + \frac{(-600)}{300}$$

$$\text{or} \quad \sum \frac{H}{T} = 0.$$

This example shows Σ H/T = 0, only in the limiting case and in no case Σ H/T is greater than zero.

1.42 ENTROPY AND THE SECOND LAW OF THERMODYNAMICS

Consider a closed system undergoing a reversible process from state 1 to state 2 along the path A and from state 2 to state 1 along the path B (Fig. 1.24). As this is a reversible cyclic process

$$\oint \frac{\delta H}{T} = 0$$

$$\therefore \quad \int_{1A}^{2A} \frac{\delta H}{T} + \int_{2B}^{1B} \frac{\delta H}{T} = 0 \qquad \text{...(i)}$$

Now, consider the reversible cycle from state 1 to state 2 along the path A and from state 2 to state 1 along the path 0.

For this reversible cyclic process

$$\int_{1A}^{2A} \frac{\delta H}{T} + \int_{2C}^{1C} \frac{\delta H}{T} = 0 \qquad \text{...(ii)}$$

From equations (i) and (ii)

$$\int_{2B}^{1B} \frac{\delta H}{T} + \int_{2C}^{1C} \frac{\delta H}{T} \qquad \text{...(iii)}$$

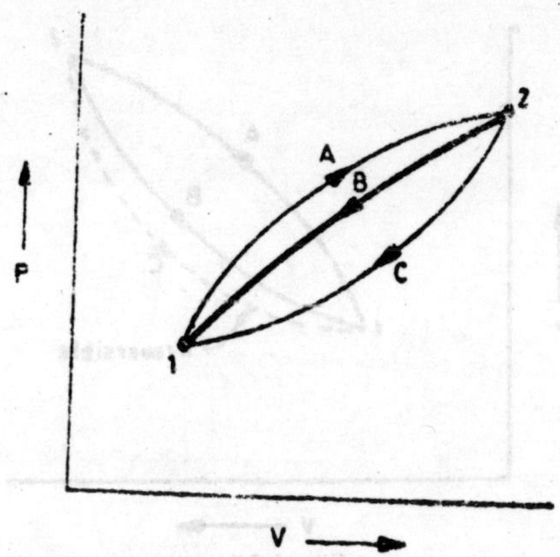

Fig. 1.24

This shows that $\int \delta H/T$ has the same value for all the reversible paths from state 2 to state 1. The quantity $\int \delta H/T$ is independent of the path and is a function of the end states only, therefore it is a property.

This property is called entropy. Entropy is a thermodynamical property and is defined by the relation

$$dS = \frac{\delta H}{T} \quad \text{...(iv)}$$

or $$S_2 - S_1 = \int_1^2 \frac{\delta H}{T} \quad \text{...(v)}$$

The quantity $S_2 - S_1$ represents the change in entropy of the system when it is changed from state 1 to state 2.

1.43 ENTROPY CHANGES OF A CLONED SYSTEM DURING AN IRREVERSIBLE PROCESS

Consider a reversible cycle where the state is' changed from 1 to 2 along the path A and 2 to 1 along the path B (Fig. 1.25).

For a reversible cyclic process

$$\oint \delta H = 0$$

$$\int_{1A}^{2A} \frac{\delta H}{T} + \int_{2B}^{1B} \frac{\delta H}{T} = 0$$

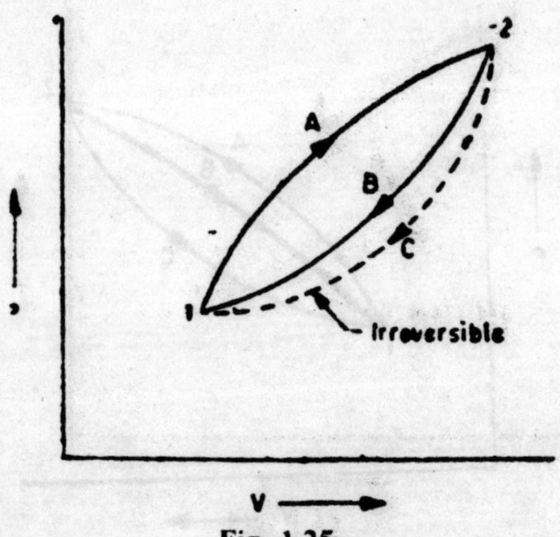

Fig. 1.25

Now, consider an irreversible path C, from state 2 to state 1. Applying Clausius inequality for the cycle of processes A and C,

$$\oint \frac{\delta H}{T} < 0$$

$$\therefore \qquad \int_{1A}^{2A} \frac{\delta H}{T} + \int_{2C}^{1C} \frac{\delta H}{T} < 0 \qquad \text{...(ii)}$$

From equations (i) and (ii)

$$\int_{2B}^{1B} \frac{\delta H}{T} - \int_{2C}^{1C} \frac{\delta H}{T} > 0$$

Since path B is reversible and entropy is a property

$$\int_{2B}^{1B} \frac{\delta H}{T} = \int_{2B}^{1B} dS = \int_{2C}^{1C} dS$$

$$\therefore \qquad dS > \frac{\delta H}{T} \qquad \text{...(iii)}$$

$$\text{or} \qquad S_2 - S_1 > \int_1^2 \delta H \qquad \text{...(iv)}$$

To conclude,

For a reversible process

$$S_2 - S_1 > \int_1^2 \frac{\delta H}{T}$$

Equation (iv) show that the effect of irreversibility is always to increase the entropy of a system.

1.44 ENTROPY

Consider adiabatics L and M on the P – V indicator diagram (Fig. 1.27). All along the adiabatic L, with change in pressure there is change in volume and temperature. This shows that all along the adiabatics L or M, there is change of temperature. Consider the isothermals at temperatures T_1, T_2 and T_3. ABCD represents the Carnot's reversible cycle. From A to B, heat energy H_1 is absorbed at temperature T_1. From C to D, heat energy H_2 is rejected at temperature T_2.

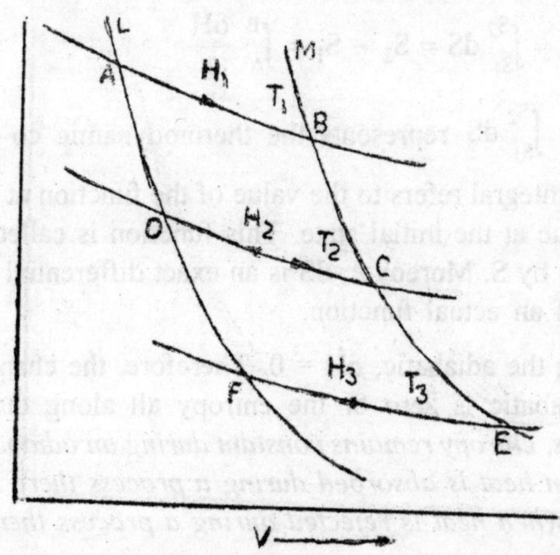

Fig. 1.26

$$\frac{H_1}{T_1} = \frac{H_2}{T_2}$$

Similarly considering the cycle DCEF

$$\frac{H_2}{T_2} = \frac{H_3}{T_3}$$

$$\therefore \quad \frac{H_1}{T_1} = \frac{H_2}{T_2} = \frac{H_3}{T_3} = \text{constant}$$

From one adiabatic to the other adiabatic, heat energy is either absorbed or rejected. The quantity of heat absorbed or rejected is not constant but it depends upon the temperature. Higher the temperature, more is the heat energy absorbed or rejected and *vice versa*. The quantity H/T between two adiabatics is constant and this is called the change in entropy. Let the entropy for the adiabatics L and M be S_1 and S_2 respectively.

Here S_1 and S_2 are arbitrary quantities.

$$S_2 - S_2 = \frac{H}{T} \text{ constant}$$

If the adiabatics are very close, and the heat absorbed or rejected is δH at a temperature T,

Change in entropy

$$dS = \frac{\delta H}{T} \qquad ...(1)$$

In general, the change in entropy

$$= \int_{S_1}^{S_2} dS = S_2 - S_1 = \int_A^B \frac{\delta H}{T} \qquad ...(2)$$

$\int_A^B \frac{\delta H}{T} = \int_{S_1}^{S_2} dS$ represents the thermodynamic co-ordinate of a system. This integral refers to the value of the function at the final state minus its value at the initial state. This function is called entropy and is represented by S. Moreover, dS is an exact differential since it is the differential of an actual function.

All along the adiabatic, $\delta H = 0$. Therefore, the change in entropy along an adiabatic is zero or the entropy all along the adiabatic is constant. *Thus, entropy remains constant during an adiabatic reversible process. When heat is absorbed during a process there is increase in entropy and when heat is rejected during a process there is decrease in entropy.*

1.45 CHANGE IN ENTROPY IN A REVERSIBLE PROCESS (CARNOT'S CYCLE)

Consider a complete reversible process [Garnet's cycle] ABCDA (Fig. 1.27). From A to B, heat energy H_1 is absorbed by the working substance at temperature T_1. The gain in entropy of the working substance from A to B = H_1/T_1. (H_1/T_1 is the decrease in entropy of the source from which the amount of heat H_1 is drawn at a temperature T_1). From B to C there is no change in entropy because BC is an adiabatic. From C to D, heat energy H_2 is rejected by the working substance at a temperature T_2. The loss in entropy of the working substance from C to D = H_2/T_2. (H_2/T_2 is also the gain in entropy of the sink to which the amount of heat H_2 is rejected at a temperature T_2). From D to A there is no change in entropy. Thus, the total gain in entropy by the working substance in the cycle ABCDA

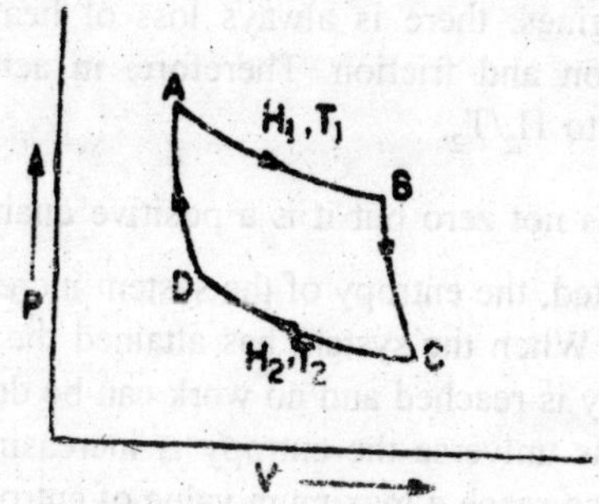

Fig. 1.27

$$= \frac{H_1}{T_1} \frac{H_2}{T_2}$$

But for a complete reversible process

$$\frac{H_1}{T_1} = \frac{H_2}{T_2}$$

Hence the total change in entropy of the working substance in a complete reversible process

$$= \oint dS = \frac{H_1}{T_1} - \frac{H_1}{T_2} = 0$$

1.46 CHANGE IN ENTROPY IN AN IRREVERSIBLE PROCESS

In an irreversible process like conduction or radiation, heat is lost by a body at a higher temperature T_1 and is gained by the body at a lower temperature T_2. Here T_1 is greater than T_2.

Let the quantity of heat given out by a body at a temperature T_1 be B and the heat gained by the body at a temperature T_2 be B. Consider the hot and the cold bodies as one system.

Loss in entropy of the hot body $= H/T_1$

Gain in entropy of the cold body $= H/T_2$

Therefore, the total increase in entropy of the system

$$= \frac{H}{T_2} - \frac{H}{T_1}$$

It is a positive quantity because T_2 is less than T_1. Thus the entropy of the system increases in all irreversible processes.

1.47 THIRD LAW OF THERMODYNAMICS

In all heat engines, there is always loss of heat in the form of conduction, radiation and friction. Therefore, in actual heat engines H_1/T_1 is not equal to H_2/T_2.

$\therefore \frac{H_1}{T_1} - \frac{H_2}{T_2}$ is not zero but it is a positive quantity. When cycle after cycle is repeated, the entropy of the system increases and tends to a maximum value. When the system has attained the maximum value, a stage of stagnancy is reached and no work can be done by the engine at this stage. In this universe the entropy is increasing and ultimately the universe will also reach a maximum value of entropy when no work will be possible. With the increase in entropy, the disorder of the molecules of a substance increases. The entropy is also a measure of the disorder of the system. With decrease in entropy, the disorder decreases. *At absolute zero temperature, the entropy tends to zero and the molecules of a substance or a system are in perfect order (well arranged). This is the third law of thermodynamics.*

Example:

The molecules are more free to move in the gaseous, state than in the liquid state. The entropy is more in the gaseous state than in the liquid state. The molecules are more free to move in the liquid state than in the solid state. The entropy is more in the liquid state than in the solid. Thus when a substance is converted from a solid to a liquid and then from the liquid to the solid state, the entropy increases and *vice versa.* When ice is converted into water and then into steam, the entropy and disorder of the molecules increase. When steam is converted into water and then into ice, the entropy and disorder of the molecules decrease. *Hence entropy is a measure of the disorder of the molecules of the system.*

By any ideal procedure, it is impossible to bring any system to absolute zero temperature performing a finite number of operations. This is called the principle of unattainability of absolute zero. Thus according to Fowler and Guggenheim, the unattainability principle is called the third law of thermodynamics.

1.48 TEMPERATURE ENTROPY DIAGRAM

The temperature-entropy diagram is used in engineering and meteorology. Consider the Carnot's cycle ABCDA [Fig. 1.28 (i)]. From

A to B, heat energy H_1 is absorbed at temperature T_1. The- increase in entropy S_1, takes place from A to B [Fig. 1.28 (ii)]. From B to C, there is no change in entropy. The temperature decreases at constant entropy. From C to D, there is decrease in entropy (S_2) at constant temperature T_2. From D to A, there is no change in entropy but the temperature increases.

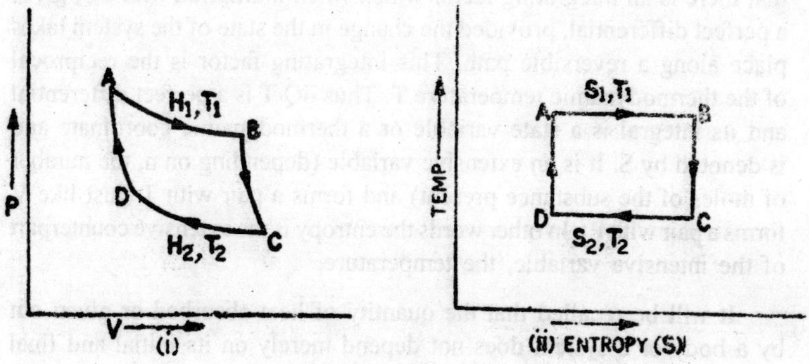

Fig. 1.28

The area ABCD in the temperature-entropy diagram represents the actual amount of energy converted into work [Fig. 1.28 (ii)].

The area ABCD = $S_1 (T_1 - T_2) = S_2 (T_1 - T_2)$

But $\quad S_1 - \dfrac{H_1}{T_1}$ and $s_2 \dfrac{H_2}{T_2}$

Here $\quad S_1 = S_2 = \dfrac{H_1}{T_1} = \dfrac{H_2}{T_2} = \dfrac{H_1 - H_2}{T_1 - T_2}$

$\therefore$ Area $\quad$ ABCD $= \dfrac{(H_1 - H_2)}{T_1 - T_2} = H_1 = 1 - \dfrac{T_2}{T_1}$

Therefore, the area ABCD represents the energy converted to work

Efficiency $\quad = \dfrac{H_1 - H_2}{H_1} = -1 \dfrac{H_2}{H_1} = 1 - \dfrac{T_2}{T_1}$

Here H_2 is the unavailable energy.

$$H_2 = \frac{H_1}{T_1} \times T_2 = S_1 \times T_1$$

The unavailable energy depends on the change in entropy at temperature T_1 and the temperature T_2.

1.49 ENTROPY

Clausius felt that a thermodynamic systems needed one more physical variable in addition to pressure?, volume V, temperature T and internal energy U, to describe completely its state. This entity, the entropy, had also a fixed value for a thermodynamic system. He showed (in 1854) that there is an integrating factor which when multiplied with δQ, gives a perfect differential, provided the change in the state of the system takes place along a reversible path. This integrating factor is the reciprocal of the thermodynamic temperature T. Thus $\delta Q/T$ is a perfect differential and its integral is a state variable or a thermodynamic coordinate and is denoted by S. It is an extensive variable (depending on n, the number of moles of the substance present) and forms a pair with T (just like V forms a pair with P). In other words the entropy is the extensive counterpart of the intensive variable, the temperature,

It will be recalled that the quantity of heat absorbed or given out by a body or a system does not depend merely on its initial and final state but also on the manner in which it has passed from its initial state to final state (*i.e.*, on the path along which it passed). Whence the quantity of heat exchanged during a small step in the process is not a perfect differential hence it is to be represented by δQ (not δQ). But the differential $\delta Q/T$ gives the change in the entropy, dS of a homogeneous system and is a point function. The product TdS has .the dimensions of energy and it measures the quantity of heat given to the body, (or system) at temperature Tin a quasistatic reversible process, to bring about the change in dS, in the entropy of the system.

Now, although this new quantity entropy is quite as real a physical quantity as volume, pressure and temperature yet it has something intangible about it, for there is nothing physical to represent it. It has therefore been called a ghostly quantity. But as some one has rightly remarked, 'the more shadowy the conception to be visualised, the greater the need of a material anology'. Looking, therefore for a suitable analogy, the nearest to it, we can think of is mass or inertia. We can therefore imagine entropy to be thermal inertia, which bears a relation to heat motion similar to the one that mass and moment of inertia bear to translatory and rotatory motion respectively. The meaning and significance of the term entropy will become increasingly clearer as we proceed along.

1.50 ENTROPY AND ADIABATICS

Let A_1 and A_2 be two adiabatics on a P-V diagram as shown in (Fig. 1.29). Then, as we know, there is a regular change of temperature (accompanied by change in pressure and volume) along them. Let the isothermals AB, CD, EF... correspond to temperatures T_1, T_2, T_3..respectively be drawn across the two adiabatics as shown. Now we have four sided figures ABDC, CDFE etc.) each representing a Carnot cycle operating between temperatures T_1, T_2; T_2, T_3, etc.

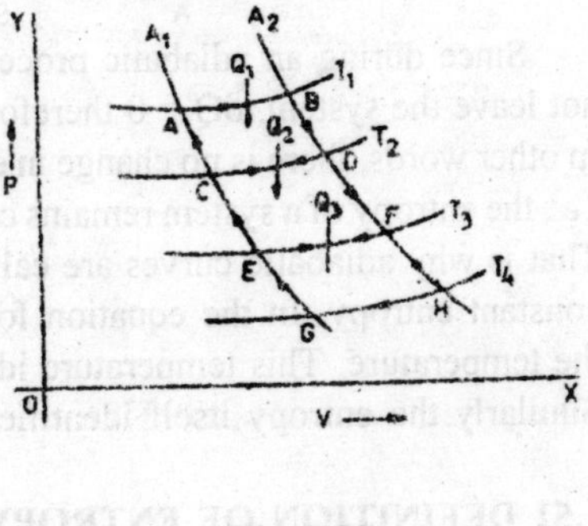

Consider first cycle ABDC If Q_1 be the quantity of heat absorbed at temperature T_1 along the isothermal AB and Q_2, the quantity of heat rejected at temperature T_2 along the isothermal DC we have

Fig. 1.29

$$\frac{Q_1}{T_1} = \frac{Q_2}{T_2}$$

Similarly, considering another cycle CDFE, the quantity of heat absorbed at temperature T_1 along the isothermal CD is Q_2 and the quantity of heat rejected at temperature T_3 along the isothermal FE be Q_3, we have

$$\frac{Q_2}{T_2} = \frac{Q_3}{T_3}$$

and for vet another cycle

$$\frac{Q_3}{T_3} = \frac{Q_4}{T_4} \text{ etc.}$$

These equations read together suggest

$$\frac{Q_1}{T_1} = \frac{Q_2}{T_2} = \frac{Q_3}{T_3} = \frac{Q}{T} = \text{constant}$$

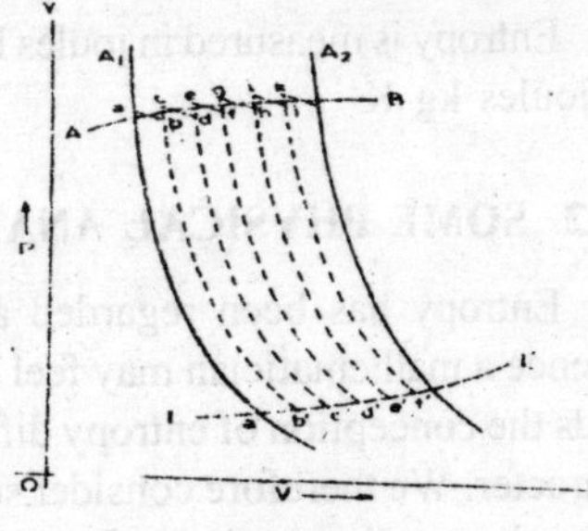

Fig. 1.30

This shows that the quantity Q/T remains constant during transition between the two adiabatics along any one of the isothermals chosen. Here, as we have seen, Q is the quantity of heat absorbed or rejected at temperature T corresponding to that isothermal. Now if the adiabatics are close enough, so that the heat absorbed or liberated at temperature T of the isothermal, be δQ then $dS = \delta Q/T$ is change in entropy. It follows, therefore, that the change in entropy $S_2 - S_1$, during a finite change from the state of the system represented by A to that represented by B is

$$S_2 - S_1 = \int_A^B \delta\, Q/T = \int_{S_1}^{S_2} dS$$

Since during an adiabatic process, beat is neither allowed to enter not leave the system, $\delta Q = 0$ therefore the change in entropy $\delta Q/T = 0$. In other words, there is no change in entropy during an adiabatic process, *i.e.*, the entropy of a system remains constant during an adiabatic process. That is why adiabatic curves are called the isentropics or the curves of constant entropy. In the equation for entropy we have one more term the temperature. This temperature identifies the rank of, an isothermal. Similarly the entropy itself identifies the rank of an adiabatic.

1.51 DEFINITION OF ENTROPY

In the above discussion we have considered the transition from one adiabatic to another (or from state A to state B) along an isothermal. This was, however, not necessary. All that is needed is that the change should occur along a reversible path. So let us consider that the change occurs along a reversible path AB as shown in (Fig. 1.31) where AB is not an isothermal.

Entropy is measured in joules K^{-1} and the specific entropy is measured in joules kg K^{-1}.

1.52 SOME PHYSICAL ANALOGIES OF ENTROPY

Entropy has been regarded as a function defined by an equation whence a mathematician may feel quite at home with it. But the physicist finds the conception of entropy difficult to grasp because of its intangible character. We therefore consider some analogies for this ghostly quantity for understanding its thermal manifestation. (Note: be always careful not to pursue analogies too far). Now from definition

$$dS \equiv \frac{\text{heat energy}}{\text{temperaturre}}$$

$$\therefore \text{heat energy} \equiv dS \times \text{temperature.}$$

Now compare it with

potential energy ≡ Mg × h

Kinetic energy ≡ $\frac{P}{2m} \times P$

electrical energy ≡ C × V

thermal energy ≡ 1/2k × T

rotational energy ≡ 1/2I × ω^2.

The quantity that follows after × is easily measurable and increase/decrease in its value gives the change in the energy level *i.e.*, h, ω, ς etc. may be taken to be level (potential) for the energy and first quantity, namely M, I may be similar to entropy. A.M. Worthington and S.G. Wheeler employed rotational motion in illustration of the properties of entropy, where rotational inertia I was not necessarily constant. It depends on arrangement of the masses. Entropy, for the purpose of analogy, was assumed to be the thermal inertia of a system. They associated it with all kinds of motion of particles of the system which measured the level of disorder.

1.53 CHANGE OF ENTROPY IN A CARNOT CYCLE

A carnot cycle consists of two isothermals and two adiabatics all of which are reversible. During the two adiabatic processes there is no change in entropy. So, changes in entropy occur only during the two isothermal processes of these, one is an isothermal expansion during which heat Q_1 is absorbed at constant temperature T_1 and the gain in entropy is therefore Q_1/T_1. The other is an isothermal compression, during which heat Q_2 is given out at a constant temperature T_2 and so the loss in entropy is Q_2/T_2. Now, since in a Carnot cycle, $Q_1/T_1 = Q_2/T_2$, it follows that the gain in entropy is equal to the loss in entropy. So the change in entropy through the cycle is zero. Mathematically,

Let us draw any arbitrary isothermal I → I' corresponding to a temperature T^1 across the two adiabatics and consider A B to be made up of a number of elements of isothermal and adiabatic curves, alternating with each other, as shown by the dotted zig-zag line. Like: ab—a small element of an isothermal curve, ab—a small element of an adiabatic curve and so on. Thus, the whole transition along AB may be supposed to be made of small transitions along these elements of isothermal and adiabatic curves.

If we produce the elementary adiabatic curves to meet the isothermal $I \rightarrow I'$, we have a number of elementary Carnot cycles like abb^1a^1, cdc^1d^1 etc. If, therefore, δQ be the quantity of heat absorbed along the isothermal ab at temperature T and δQ^1 be the quantity of heat rejected along with isothermal b^1a^1 at temperature T^1 we have $\delta Q/T = \delta Q^1/T^1$. In the limit, when Carnot cycles become infinitely small the zig-zag curve coincides with the reversible path AB and we have

$$\int_A^B \delta Q/T = \int \delta Q^1/T^1$$

Since $\delta Q^1/T^1$ represents the change in entropy between the two adiabatics A_1 and A_2 through A and B respectively, along the isothermal $I \rightarrow I'$. It must be a constant.

$$\therefore \quad \int_A^B \delta Q/T = \text{constant}$$

i.e., the change in entropy in passing from a state A to a state B is constant irrespective of the reversible path along which the change takes place. As we have seen here, the change in entropy of a system when it passes from one state A to another state B depends upon its initial and final states only and is quite independent of the manner in which the change is brought about. This is exactly the property of an actual function of state.

Hence $\delta Q/T$ is a function of the thermodynamic coordinates or state variables of the point ^ and 5 and its integral defines the change in entropy of the system during transition from state A to state B.

If the state A be arbitrarily chosen as the zero state of entropy, S2-S1 gives the entropy S of the system in state B relative to the arbitrarily chosen zero entropy state A. We may thus define entropy as a single valued function of the parameters of the state (like pressure, volume) for all those states that can be reached by a reversible process from some arbitrarily chosen initial zero state.

Since we are, in general, concerned only with changes of entropy from one state to another, we need hardly bother about the standard zero state of entropy. Quite often, however, the entropy of a substance at NTP is taken to be zero.

$$\frac{Q_1}{T_1} = -\frac{Q_2}{T_2} \text{ or, } \frac{Q_1}{T_1} + \frac{Q_2}{T_2} = 0$$

for the whole cycle (negative sign indicates loss of entropy).

Thus, during a Carnot cycle, the entropy of a system remains unchanged.

1.54 CHANGE OF ENTROPY IN ANY REVERSIBLE CYCLE

The results obtained above for a Carnot cycle can be shown to be true for any reversible cycle by a method first suggested by Clausius.

Consider a reversible cycle, the indicator diagram for which is a closed curve ABCD (Fig 1.31). This may be divided into a number of Carnot cycles by drawing across the cycle a series of adiabatics A_1, A_2, A3,...close together with short isothermals ab, cd,.,.,a^{1}b^1..introduced between them as shown. Then clearly each such elementary Carnot cycle formed like abb^{1}a^1,cdd^{1}c^1,...is bounded by two adiabatics and two isothermals. So that, for each one of these cycles

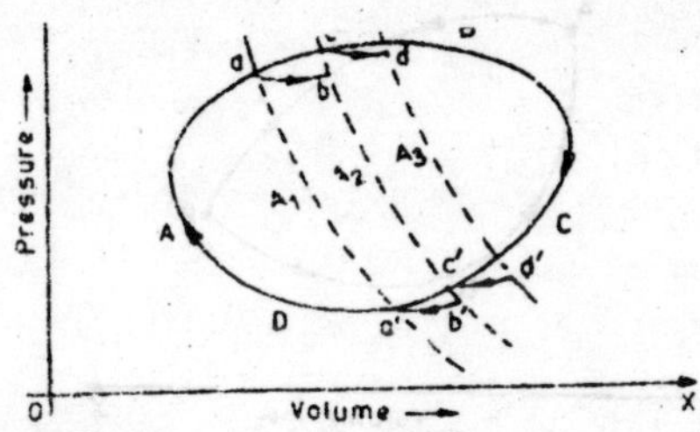

Fig. 1.31

$$\frac{\delta Q_1 i}{T_{1i}} + \frac{\delta Q_{2i}}{T_{2i}} = 0$$

where δQ_{1i} is the heat absorbed at temperature T_{1i}, δQ_{2i} is the heat rejected at temperature T_{2i}, etc.

Summing up these results for all the elementary cycles we have

$$\sum_i \left(\frac{\delta Q_{1i}}{T_{1i}} + \frac{\delta Q_{2i}}{T_{2i}} \right) = 0$$

or, if the elementary cycles be infinitely small

$$\oint \frac{\delta Q}{T} = 0 \oint dS$$

This is known as *Clausius theorem*. It states that the change in entropy for a complete reversible thermodynamic cycle is zero.

Corollary

We may use the above result for a reversible cycle to prove that "the entropy of a system depends upon its state only and is quite independent of the manner in which this state has been arrived at." For if the system (or the substance) be taken by a reversible path ACB, from the state represented by A to that represented by B. Clearly, difference between the entropies of the system at A and at B (or the change in entropy of the system in going from A to B) in (Fig 1.32) is

$$\int_A^B \frac{\delta Q}{T} = \int_{S_A}^{S_B} dS = S_B - S_A$$

where S_A and S_B are the entropies of the system at A and B respectively.

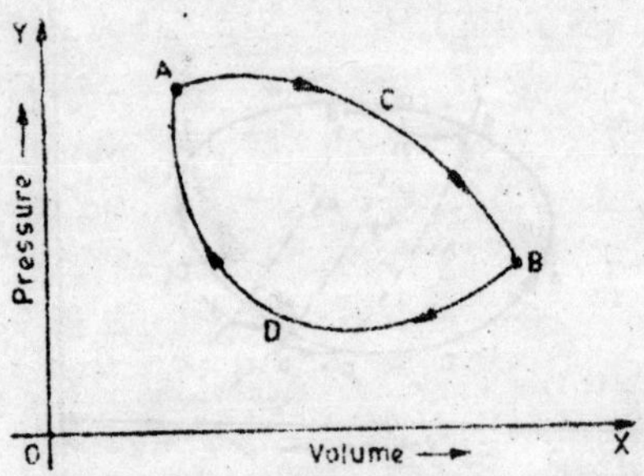

Fig. 1.32

If now the system be supposed to come back to its original state represented by A, along the path BDA, it will assume its original entropy and so the change in its entropy will be zero, *i.e.* change in entropy from A to B along ACB + change in entropy from B to A along BDA = 0, taking proper care of the signs. It follows, therefore, that the change in entropy of the system when taken from the state represented by A to other state represented by B is the same whether the path followed is ACB, ADB or any other. To state the same thing differently, we may say that all processes leading from a state A to a state B result in the same change in entropy—Entropy of the system is thus a point function on an indicator diagram. The entropy may be considered as an outcome of the Second Law of Thermodynamics—a new but equally good function of state. It touches the aspect of arrangement of particles and their motions (randomness) on the body. In fact the Second Law of Thermodynamics suggests that an ordered state tends to become disordered. "The direction which a self-sustained machine follows is that in which disorder increases."

1.55 CALCULATION OF CHANGE IN ENTROPY ALONG AN IRREVERSIBLE PATH

It has been emphasised severally that the above results apply only to a reversible process and not to an irreversible one. The expression $\int \delta Q/T$ may not give, therefore, the change in entropy in general. How then, is it possible to calculate a change in entropy if the change of state from A to B is brought about by an irreversible process? The corollary of section 9.6 gives us a clue:

We simply find some reversible process to bring the substance from state A to slate B. For this process, the change in entropy is given by $\int \delta Q/T$ and may be easily calculated. Now, since the change in entropy depends only on the initial and final states of the system, and not upon the manner of change, then entropy-change, from A to B by the irreversible process too must be the same as by the reversible process. This then, calculated as above, gives the entropy-change for the irreversible process.

Let us consider (Fig 1.32). Let the initial state of the system be A and the system follows the path AEB to attain the final state B. Furhter, let AEB be a non-reversible path. If we are to find SB-SA we may draw reversible isothermal AC and a reversible adiabetic BCN through A and B respectively. Then if the two meet at C we have, clearly, no change in entropy along BCN. Hence $S_B - S_A$ is simply S_C-S_A along the reversible isothermal AC, which is given by Q/T.

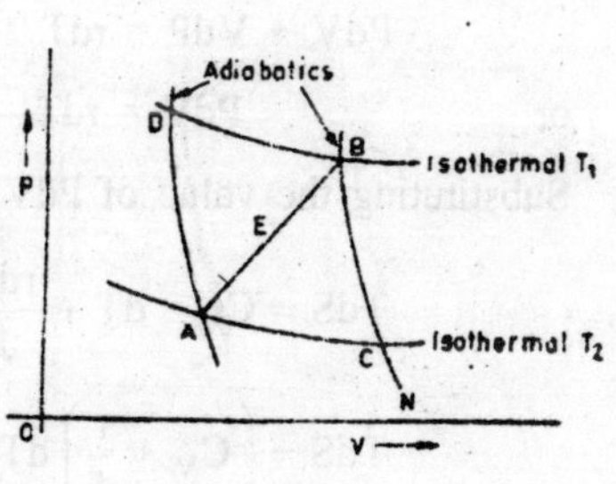

Fig. 1.33

1.56 ENTROPY OF PERFECT GAS

Consider one gram of a perfect gas at a pressure P_1 volume V and temperature T. Let the quantity of heat given to the gas be δH.

$$\delta H = dU + \delta W$$

$$\delta H + 1 \times C_V \times dT + PdV/J. \qquad ...(i)$$

$$\delta B = TdS$$

$$\therefore \quad TdS = C_V dT + PdV/J \quad \text{...(ii)}$$

Also $\quad PV = rT$

or $$P = \frac{rT}{V}$$

$$\therefore \quad TdS = C_V dT + \frac{rT.dV}{JV}$$

$$dS = C_V \frac{dT}{T} + \frac{r}{J}\frac{dV}{V}$$

Integrating, $$\int_{S_1}^{S_2} dS = C_V \int_{T_1}^{T_2} \frac{dT}{T} + \frac{r}{J}\int_{V_1}^{V_2} \frac{dV}{V}$$

$$S_2 - S_1 = C_V \log_e \frac{T_2}{T_1} + \frac{r}{T} \log_e \frac{V_2}{V_1} \quad \text{...(iii)}$$

$$S_2 - S_1 = C_V \times 2.3026 \log_{10} \frac{T_2}{T_1} + \frac{r}{T} \times 2.3026 \log_{10} V_2/V_1 \quad \text{...(iv)}$$

The change in entropy can be calculated in terms of pressure also.

$$PV = rT$$

Differentiating

$$PdV + VdP = rdT$$

or $$PdV = rdT - VdP$$

Substituting the value of PdV in equation (ii)

$$TdS = C_V \times dT + \frac{rdT}{J} - \frac{VdP}{J}$$

$$TdS = \left(C_V + \frac{r}{J}\right) dT - \frac{VdP}{J}$$

But $$C_V + \frac{r}{J} = C_P$$

$$\therefore \quad dS = C_P \frac{dT}{T} - \frac{VdP}{JT}$$

Also $\quad PV = rT$

or $$\frac{V}{T} = \frac{r}{T}$$

$$\therefore \quad dS = C_P \frac{dT}{T} - \frac{r}{J} \frac{dP}{P}$$

Integrating

$$\int_{S_1}^{S_2} dS = C_P \int_{T_1}^{T_2} \frac{dT}{T} - \frac{r}{J} \int_{P_1}^{P_2} \frac{dP}{P}$$

$$S_2 - S_1 = C_P \log_e \frac{T_2}{T_1} - \frac{r}{T} \log_e \frac{P_2}{P_1} \quad \text{...(v)}$$

$$S_2 - S_1 = C_P \times 2.3026 \times \log_{10} \frac{T_2}{T_1} - \frac{r}{T} \times 2.3026 \log_{10} \frac{P_2}{P_1} \quad \text{...(vi)}$$

Note : r is the ordinary gas constant and has to be taken in units of work, C_p represents the specific beat for I gram of a gas at constant pressure.

If C_p represents gram molecular specific heat of a gas at constant pressure and 22 the universal gas constant, then

$$S_2 - S_1 = C_P \times 2.3026 \log_{10} \frac{T_2}{T_1} - \frac{r}{T} \times 2.3026 \log_{10} \frac{P_2}{P_1} \quad \text{...(vii)}$$

1.57 ZERO POINT ENERGY

According to kinetic theory, the energy of a system at absolute zero should be zero. It means the molecules of the system do not possess any motion. But according to the modern concept, even at absolute zero, the molecules are not completely deprived of their motion and hence possess energy. The energy of the molecules at absolute zero temperature is called zero point energy.

1.58 NEGATIVE TEMPERATURES

The specific heat of a substance decreases with increase in temperature. However, the specific heat does not tend to zero as the temperature tends to infinity. This shows that the temperature has a +ve sign only.

But recent experiments by Ramsey (1956) have shown that a part of a system *i.e.*, the nucleus of a solid, can have a negative temperature. This sub-system is considered isolated from the main system (*i.e.*, solid lattice). The specific heat of the sub-system tends to zero at high temperature. A small amount of heat energy tends to raise the temperature of the system to infinity. It is possible to add still more energy to the sub-system at infinity and it forces the sub-system into the negative, temperature region. It has been shown by microscopic statistical analysis that there is no distinction between the temperature of $+\infty$ and $-\infty$. In thermodynamics, the parameter 1/T is more significant than T.

The negative temperatures are hotter than the positive temperatures and minus zero (–0) is the hottest temperature and plus zero (+0) is the coldest temperature.

The negative temperature is not possible with the system as a whole and is only an exception to the rule that only positive temperatures exist. The negative temperatures are possible only for isolable sub-systems. For all normal purposes the temperatures are always positive.

1.59 HELMHOLTZ FUNCTION

Helmholtz function F, is the property of a system and is given by the equation

$$F = U - TS \qquad \text{...(i)}$$

In practice, the primary function of heat engines and other devices is to perform mechanical work. From the first law of thermodynamics, for a system working between two equilibrium states,

$$\delta H = dU + \delta W \qquad \text{...(ii)}$$

or
$$\delta W = \delta H - dU \qquad \text{...(iii)}$$

It means that the energy converted into work is provided partly by the heat reservoir with which the system is in contact and which gives up a quantity of heat δH and partly by the system whose internal energy decreases by (–dU).

When a system undergoes a process between two equilibrium state, how much maximum work can be done by it? This can be derived by assuming that the system exchanges heat energy only with a single heat reservoir at a temperature T_0.

From the principle of increase of entropy, it is known that the sum of the increase in entropy of the system and that of the surroundings is equal to or greater than zero. Suppose,

Increase in entropy of the system = dS

Increase in entropy of the surroundings (reservoir) = dS_0

Heat absorbed by the system from the reservoir = δB

Temperature of the reservoir = T_0

$$\therefore \quad dS + dS_0 > 0 \qquad \text{...(iv)}$$

For a reservoir. $dS_0 = \dfrac{-\delta H}{T_0}$ [Since reservoir has given heat]

$$\therefore \quad dS - \frac{\delta H}{T_0} > 0$$

$$\text{or} \quad T_0 \,.\, dS > \delta H \qquad \text{...(v)}$$

From equations (iii) and (v)

$$\delta W < T_0 \,.\, dS - dU \qquad \text{...(vi)}$$

The quantities U and S are the properties of the system and T_0 is a constant. Therefore, for a finite process between state 1 to state 2, by integrating equation (vi),

$$W < T_0\,(S_2 - S_1) - (U_2 - U_1) \qquad \text{...(vii)}$$

$$W < (U_1 - U_2) - T_0(S_1 - S_2) \qquad \text{...(viii)}$$

Suppose that the initial and the final temperatures are equal and are the same as that of the heat reservoir.

$$T = T_0$$

Here T is the temperature of the system.

Equation (viii) can be written as,

$$\therefore \quad W_r < (U_1 - U_2)_T - T(S_1 - S_2)_T \qquad \text{...(ix)}$$

The Helmholtz function,

$$F = U - TS.$$

Therefore for two equilibrium states 1 and 2 at the same temperature T,

$$(F_1 - F_2) = (U_1 - TS_1) - (U_2 - TS_2)$$

$$= (U_1 - U_2) - T(S_1 - S_2)$$

$$\text{or} \quad (F_1 - F_2)_T = (U_1 - U_2)_T - T(S_1 - S_2)_T \qquad \text{...(x)}$$

From (ix) and (x)

$$W_T < (F_1 - F_2)_T \quad ...(xi)$$

Thus, the maximum work that can be done in any process between two equilibrium states at the same temperature is equal to the decrease in the Helmholtz function of the system. Here the system exchanges heat with a single heat reservoir at the same temperature. Moreover maximum work is done when the process if reversible. If the process is irreversible, the work done is less than the maximum.

It is to be remembered that the maximum work that can be done is equal to the decrease in the Helmholtz function of a system but the energy converted into work is provided partly by the system and the remaining by the heat taken from the heat reservoir.

1.60 THERMODYNAMIC POTENTIAL OR GIBBS FUNCTION

The Gibbs function G of a system is given by,

$$G = U - TS + PV.$$

Consider a system that can do other forms of work, in addition to P.dV work, *e.g.*, a voltaric cell. In the case of a voltaic cell, the electrical work is –E.dI. Similarly for a magnetic material, the magnetic work is –m.dH. In general, the work will be given by P.dV plus a sum of terms each being the product of intensive variable (such as P, E or m) and the differential of an extensive variable (such as dV, dI and dH). In the case of a voltaic cell, the intensive variable is S and differential of extensive variable is dI.

Suppose, in general, in addition to PdV the intensive variable is y and differential of extensive variable is dx. The work done for any reversible process,

$$\delta W = PdV + ydx$$

$$W = \int_{V_1}^{V_2} PdV + \int_{x_1}^{x_2} ydx$$

Take $\int_{x_1}^{x2} ydx = A$

Consider a process where the system works at constant pressure P_0 and the change in volume is $(V_2 - V_1)$

$$\therefore \quad \int_{V_1}^{V_2} PdV = P_0 [V_2 - V_1]$$

$$\therefore \quad W = P_0[V_2 - V_1] + A \quad ...(i)$$

For a system that exchanges heat with a reservoir temperature

$$W < (U_1 - U_2) - T_0 (S_1 - S_2) \quad ...(ii)$$

$$P_0[V_2 - V_1] + A < (U_1 - U_2) - T_0 (S_1 - S_2)$$

or $$A < (U_1 - U_2) - T_0 (S_1 - S_2) + P_0[V_1 - V_2] \quad ...(iii)$$

Consider a specific process, where the initial and the final states of the system and the surroundings are at the same temperature (T_0) and pressure (P_0)

$$To = T \text{ and } P_0 = P.$$

From equation (iii)

$$A_{P,T} < (U_1 - U_2)_{P,T} - T(S_1 - S_2)_{P,T} + P[V_1 - V_2]_{P,T} \quad ...(iv)$$

But, for the Gibbs function,

$$G = U - TS + PV. \quad ...(v)$$

Therefore, for two equilibrium states at the same pressure and temperature,

$$[G_1 - G_2)_{P,T} = (U_1 - U_2)_{P,T} - T[S_1 - S_2]_{P,T} + P[V_1 - V_2]_{P,T} \quad ...(vi)$$

From equations (iv) and (vi)

$$A_{P,T} < (G_1 - G_2)_{P,T} \quad (vii)$$

Thus the difference between Gribbs function of a system between two equilibrium states sets the maximum limit to the work in addition to PdV work, provided the initial and the final states are at the same pressure and temperature and the system exchanges heat with a single heat reservoir. The work done will be maximum when the process is reversible. The process in this case will be isothermal isobaric. If the process is irreversible, work done will be less than the maximum.

1.61 ENTHALPY

Enthalpy in an extensive thermodynamic property and is given the symbol A. Enthalpy is given by the equation

$$h = U + PV. \quad ...(i)$$

Enthalpy is defined as the sum of the internal energy and the product of pressure and volume.

To study the properties of this function, assume that the system undergoes an infinitesimal process from an initial equilibrium state to a final equilibrium state.

From equation (i)

$$dh = dU + PdV + VdP \quad ...(ii)$$

Also $\delta H = dU + PdV \quad ...(iii)$

$$\therefore \quad dh = \delta H + VdP \quad ...(iv)$$

Dividing by dT

$$\frac{dh}{dT} = \frac{\delta H}{dT} + V\frac{dP}{dT} \quad ...(V)$$

At constant pressure (isobaric process),

$$\left(\frac{\partial h}{\partial T}\right)_P = \left(\frac{\delta H}{dT}\right)_P$$

But $\left(\frac{\delta H}{dT}\right)_P = C_P$

$$\therefore \quad \left(\frac{\partial h}{\partial T}\right)_P = C_P \quad ...(vi)$$

From equation (iv), during isobaric process, dP = 0

$$\therefore \quad dh = \delta H$$

or $h_f - h_i - H \quad ...(vii)$

Therefore change in enthalpy during an isobaric process is equal to the amount of heat transferred during the process.

In a throttling process, applying first law of thermodynamics

$$H = U_f - U_i + W$$

Here $H = 0$

and $W = P_fV_f - P_oV_i$

$$\therefore \quad 0 = U_f - U_i + P_fV_f - P_iV_i$$

or $U_i + P_iV_i = U_f + P_fV_f$

or $h_i = h_f$. ...(viii)

Therefore in a throttling process, there in no change in enthalpy of the system. The initial and the final enthalpies are equal.

1.62 C_P, C_V AND μ

The specific heat at constant pressure C_p, the specific heat at constant volume C_V and the Joule-Kelvin coefficient μ are defined as follows:

$$C_P = \left(\frac{\partial h}{\partial T}\right)_P \quad \text{...(i)}$$

$$C_V = \left(\frac{\partial U}{\partial T}\right)_V \quad \text{....(ii)}$$

$$\mu = \left(\frac{\partial T}{\partial P}\right)_h \quad \text{...(iii)}$$

These three quantities are defined in terms of the thermodynamic properties viz. pressure, volume, temperature, internal energy and enthalpy. Hence, C_P, C_V and μ are also thermodynamic properties of a substance.

1.63 MAXWELL'S EQUATIONS (ALTERNATIVE METHOD)

The properties of pure substances can conveniently be represented in terms of the four functions: internal energy, enthalpy, Helmholtz function and Gibb's function.

Internal energy $= U$

Enthalpy, $A = U + PV$

Helmholtz function

$$F = U - TS$$

Gibb's function $0 = h - TS$

or $G = U - TS + PV$

All the above four quantities can be regarded as functions of

P, V, T, and S

Now, consider a system undergoing an infinitesimal reversible process from one equilibrium state to another.

(1) The internal energy changes by an amount

$$dU = \delta H - PdV$$

or $$dU + TdS - PdV \quad \text{...(i)}$$

Here U, T and P are supposed to be functions of S and V

(2) The enthalpy changes by an amount

$$dh = dU + PdV + VdP$$

$$dh = (TdS - PdV) + P.dV + VdP$$

$$dh = TdS - VdP \qquad \text{...(ii)}$$

Here h, T and V are supposed to be functions of S and P.

(3) The Helmholtz function changes by an amount

$$dF = dU - TdS - SdT$$

or $$dF = (TdS - PdV) - TdS - SdT$$

or $$dF = -SdT - PdV \qquad \text{(iii)}$$

Here F, S and P are supposed to be functions of T and V.

(4) The Gibbs function changes by an amount

$$dG = dh - TdS - SdT$$

$$dG = (TdS + VdP) - TdS - SdT$$

$$dG = -SdT + VdP \qquad \text{...(iv)}$$

Here G, S and 7 are supposed to be functions of T and P.

As U, h, F and G are actual functions, their differentials are exact.

For exact differentials

$$dZ = Mdx + Ndy \qquad \text{...(v)}$$

Here Z, M and N are all functions of x and y.

$$\therefore \quad \left(\frac{\partial M}{\partial y}\right)_x = \left(\frac{\partial N}{\partial x}\right)_y \qquad \text{...(vi)}$$

To derive the Maxwell's equations, equation (vi) can be applied.

(1) From equation (i)

$$dU = TdS - PdV$$

Here $M = T, N = P, x = S, y = V$

From equation (vi)

$$\left(\frac{\partial T}{\partial V}\right)_S = -\left(\frac{\partial P}{\partial S}\right)_V \qquad \text{...(vii)}$$

(2) From equation (ii)

$$dh = TdS + VdP$$

Here $M = T, N =V, x = S, y = P$

From equation (vi)

$$\left(\frac{\partial T}{\partial P}\right)_S = \left(\frac{\partial V}{\partial S}\right)_P \quad \text{...(viii)}$$

(3) From equation (iii)

dF + SdT – PdV

Here M = S, N = P, x =T, y = V

From equation (vi)

$$\left(\frac{\partial S}{\partial V}\right)_T = \left(\frac{\partial P}{\partial T}\right)_V \quad \text{...(ix)}$$

(4) From equation (iv)

dG = –SdT + FdP

Here M = S, – N = V, x = T, y = P

From equation (vi)

$$\left(\frac{\partial S}{\partial P}\right)_T = \left(\frac{\partial V}{\partial T}\right)_P \quad \text{...(x)}$$

Note: The equations for dU, dh, dF and dG can be obtained by remembering the sentence, 'Good Physicists have Studied Under Very Fine Teachers'. In this sentence, the letters, G, P, h, S, U, V, F, and T are to be used.

The diagram to be used is shown in Fig. 1.34.

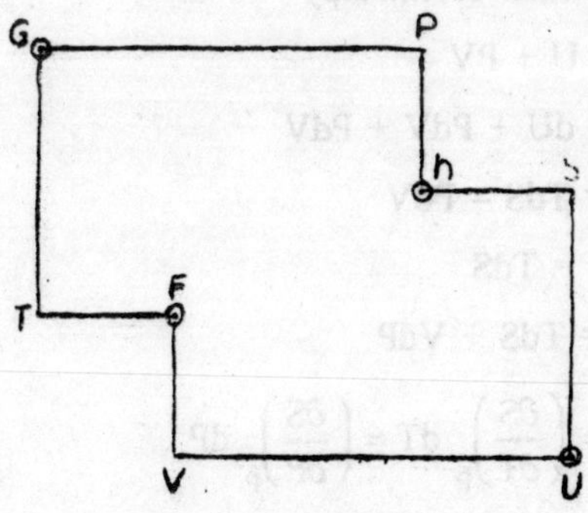

Fig. 1.34

To obtain the expression for dU, note the quantities on the two sides of U, viz. they are 8 and V. S is on the +ve direction of y-axis and V is on the –ve direction of x-axis.

$$\therefore \quad dU = +(\quad)\, dS - (\quad)\, dV$$

The quantity associated with dS is T and with dV is P.

$$\therefore \quad dU = TdS - PdV \quad \text{...(i)}$$

Similarly for dh

$$dh = (\quad)\, dS + (\quad)\, dP$$

$$dh = TdS + VdP \quad \text{...(ii)}$$

For dF

$$dF = -(\quad)\, dT - (\quad)\, dV$$

$$dF = -SdT - PdV \quad \text{...(iii)}$$

For dG

$$dG = -(\quad)\, dT + (\,)\, dP$$

$$dG + -SdT + VdP \quad \text{...(iv)}$$

1.64 JOULE-KELVIN COEFFICIENT

The Joule-Kelvin coefficient is defined as the slope of the isenthalpic curve at any point on the temperature pressure curve.

$$\mu = \left(\frac{\partial T}{\partial P}\right)_h \quad \text{...(i)}$$

From the definition of enthalpy

$$h = U + PV \quad \text{...(ii)}$$

$$\text{or} \quad dh = dU + PdV + PdV \quad \text{...(iii)}$$

$$\text{Also} \quad dU = TdS - PdV$$

$$\therefore \quad dP + PdV = TdS$$

$$\therefore \quad dh = TdS + VdP \quad \text{...(iv)}$$

$$\text{Also} \quad dS = \left(\frac{\partial S}{\partial T}\right)_P dT = \left(\frac{\partial S}{\partial P}\right)_P dP$$

$$dh = T\left(\frac{\partial S}{\partial T}\right)_P dT = \left[T\left(\frac{\partial S}{\partial P}\right)_T + V\right] dP \quad \text{...(v)}$$

The volume coefficient a is given by

But $$T\left(\frac{\partial S}{\partial T}\right)_P = C_P \qquad \text{...(vi)}$$

The volume coefficient α is given by

$$\alpha = \frac{1}{V}\left[\frac{\partial V}{\partial T}\right]_P$$

But $$\left[\frac{\partial V}{\partial T}\right]_P = -\left[\frac{\partial S}{\partial P}\right]_T = \alpha V$$

$$\therefore \quad \left[\frac{\partial S}{\partial P}\right]_T = -\alpha V \qquad \text{...(vii)}$$

Substituting these values in equation (v),

$$dh = C_P dT + \left[V - T\left(\frac{\partial V}{\partial T}\right)_P\right] dP \qquad \text{...(viii)}$$

$$dh = \frac{1}{C_P} dh + \frac{1}{C_P}\left[T\left(\frac{\partial V}{\partial T}\right)_P - V\right] dP \qquad \text{...(ix)}$$

But $$dT = \left(\frac{\partial T}{\partial h}\right)_P dh + \left(\frac{\partial T}{\partial P}\right)_h dP \qquad \text{...(x)}$$

Comparing the coefficients of equations (iv) and (x),

$$\left(\frac{\partial T}{\partial h}\right)_h = \frac{1}{C_P}\left[T\left(\frac{\partial V}{\partial T}\right)_P - V\right]$$

But $$\mu = \left(\frac{\partial T}{\partial P}\right)_h$$

$$\therefore \quad \mu = \frac{1}{C_P}\left[T\left(\frac{\partial V}{\partial T}\right)_P - V\right]$$

$$\mu C_P = \left[T\left(\frac{\partial V}{\partial T}\right)_P - V\right] \qquad \text{...(xi)}$$

$$\mu C_P = T^2 \frac{\partial}{\partial T}\left(\frac{V}{T}\right)_P \qquad \text{...(xii)}$$

From equation (viii),

$$\left(\frac{\partial h}{\partial P}\right)_T = V - T\left(\frac{\partial V}{\partial T}\right)_P \quad ...(xiii)$$

From equations (xi) and (xiii)

$$\mu C_P = -\left(\frac{\partial h}{\partial P}\right)_T \quad ...(xiv)$$

The Joule-Kelvin coefficient,

$$\mu = \frac{1}{C_P}\left[T\left(\frac{\partial V}{\partial T}\right)_P - V\right] \quad ...(xv)$$

$$\mu = \frac{V}{C_P}\left[T\left\{\frac{1}{V}\left(\frac{\partial V}{\partial T}\right)_P\right\} - 1\right] \quad ...(xvi)$$

or $$\mu = \frac{V}{C_P}[\alpha T - 1] \quad ...(xvii)$$

For a Van der Waals gas

$$\left(P + \frac{a}{V^2}\right)(V - b) = RT$$

$$PV - Pb + \frac{a}{V} - \frac{ab}{V^2} = RT$$

Differentiating, keeping P constant,

$$PdV - \frac{adV}{V^2} + \frac{2ab\,dV}{V^3} = RdT$$

or $$\frac{dV}{dT} = \frac{R}{P - \frac{a}{V^2} + \frac{2ab}{V^3}}$$

$$\left(\frac{\partial V}{\partial T}\right)_P = \frac{R}{\frac{RT}{V-b} - \frac{a}{V^2} - \frac{a}{V^2} + \frac{2ab}{V^3}}$$

$$\left(\frac{\partial V}{\partial T}\right)_P = \frac{RV^3(V-b)}{RTV^3 - 2a(V-b)^2} \quad ...(xviii)$$

Substituting this value in equation (xvi)

$$\mu = \frac{V}{C_P}\left[\frac{RTV^2\,(V-b)}{RTV^2 - 2a(V-b)^2} - 1\right]$$

$$= \frac{V}{C_P}\left[\frac{RTV^2\,(V-b) - RTV^3 + 2a(V-b)^2}{RTV^3 - 2a(V-b)^2}\right]$$

$$= \frac{V}{C_P}\left[\frac{2a\,(V-b)^2 - RTV^2b}{RTV^3 - 2a(V-b)^2}\right] \qquad \text{...(xix)}$$

From equation (xvii), as 7 and C_p are always positive, the value of μ and hence the temperature of a Van der Waals gas passing through a porous plug will depend on the value of $\alpha T - 1$.

(1) If $\alpha T > 1$, the temperature of the gas will increase.

(2) If $\alpha T < 1$, the temperature of the gas will decrease.

(3) If $\alpha T = 1$, the temperature will not change.

This temperature at which the inversion takes place is called the temperature of inversion T_i. In this case, from equation (xvii)

$\mu = 0$ and from equation (xix)

$$2a(V_{\bullet} - b)^2 = RT_iV^2b$$

$$\therefore \qquad T_i = \frac{2a(V-b)^2}{RV^2b}$$

If $V >> b$

$$T_i = \frac{2a}{Rb} \qquad \text{...(xxi)}$$

Ideal Gas

From equation (xii)

$$\mu C_P = T^2\left[\frac{\partial}{\partial T}\left(\frac{V}{T}\right)_P\right]$$

$$\therefore \qquad \mu = \frac{T^2}{C_P}\left[\frac{\partial}{\partial T}\left(\frac{V}{T}\right)_P\right]$$

For an ideal gas $\frac{V}{T} = \frac{R}{P}$

$$\therefore \qquad \mu = \frac{T^2}{C_P} \cdot \frac{\partial}{\partial T}\left[\frac{R}{P}\right]_P = 0$$

Thus, the Joule-Kelvin coefficient for an ideal gas is zero and the ideal gas passing through the porous gas does not show any change in temperature.

1.65 EQUILIBRIUM BETWEEN LIQUID AND ITS VAPOUR

Applying Gibbs potential (G), it is possible to investigate the equilibrium between a liquid and its vapour or between any two phases of a substance. Consider a closed system containing a liquid in equilibrium with its saturated vapour. The temperature and pressure are equal in both the phases. As each of the phases is in equilibrium, the temperature and pressure must remain constant throughout the phase and hence the thermodynamical coordinates F, S, U and G will be equal to the product of the specific value and the mass of the substance in that phase.

Suppose m_1 and m_2 are the masses in the liquid and vapour phases and g_1 and g_2 are the specific values of the Gibbs potential in the two phases. Then, for the whole system

$$G = m_1 g_1 + m_2 g_2 \qquad \text{...(i)}$$

If a small quantity of liquid changes into vapour, differentiating equation (i)

$$\delta G = \delta m_1 g_1 + \delta m_2 \, g_2 \qquad \text{...(ii)}$$

As the change takes place at constant temperature and pressure, the process is isothermal and isobaric

$$\therefore \qquad \delta G = 0$$

From equation (ii)

$$\delta m_1 g_1 = \delta m_2 g_2 = 0$$

But $\qquad \delta m_1 = \delta m_2$

$$\therefore \qquad g_1 = g_2$$

This shows that the thermodynamical potential per unit mass will be equal in the two phases. Equation (iii) is applicable to the processes of evaporation, fusion and sublimation.

1.66 FIRST ORDER PHASE TRANSITIONS

Consider an enclosure containing a liquid and its satur ated vapour in equilibrium. If this system undergoes an isother mal, isobaric change, then

$$g_1 = g_2 \qquad ...(i)$$

Let the temperature of the system be increased from T to T + dT. For equilibrium

$$g_1 + dg_1 = g_2 + dg_2 \qquad ...(ii)$$

or

$$dg_1 = dg_2 \qquad ...(iii)$$

If the condition of saturation is satisfied,

$$\left(\frac{dg_1}{dT}\right)_{sat.} = \left(\frac{dg_2}{dT}\right)_{sat.} \qquad ...(iv)$$

The pressure also changes from p to p + dP,

$$\therefore \quad dg_1 = \left(\frac{\partial g_1}{\partial T}\right)_P dT + \left(\frac{\partial g_1}{\partial T}\right)_T dP \qquad ...(v)$$

or

$$\left(\frac{\partial g_1}{\partial T}\right) = \left(\frac{\partial g_1}{\partial T}\right)_P + \left(\frac{\partial g_1}{\partial P}\right)_T \left(\frac{\partial P}{\partial T}\right) \qquad ...(vi)$$

But, for a unit mass,

$$dg = VdP - SdT \qquad ...(vii)$$

$$\therefore \quad \left(\frac{\partial g}{\partial P}\right)_T = V$$

and

$$\left(\frac{\partial g}{\partial T}\right)_P = -S$$

Substituting these values in equation (vi)

$$\left(\frac{\partial g_1}{\partial T}\right)_{sat.} = -S_1 + V_1 \left(\frac{\partial P}{\partial T}\right)_{sat.}$$

Similarly,

$$\left(\frac{\partial g_2}{\partial T}\right)_{sat.} = -S_2 + V_2 \left(\frac{\partial P}{\partial T}\right)_{sat.}$$

Substituting these values is equation (iv)

$$-S_1 + V_1\left(\frac{dP}{dT}\right) = -S_2 + V_2\left(\frac{dP}{dT}\right)$$

or $$\left(\frac{dP}{dT}\right)_{sat.} = \left(\frac{S_2 - S_1}{V_2 - V_1}\right) \quad \text{...(viii)}$$

But $$S_2 - S_1 = \frac{\delta H}{T} = \frac{L}{T}$$

Here L is latent heat of vaporization

$$\therefore \quad \left(\frac{dP}{dT}\right) = \frac{L}{T(V_2 - V_1)} \quad \text{...(ix)}$$

This is Clausius-Clapeyron latent heat equation.

This equation holds good in the first order phase transitions. In these processes there is transference of heat and hence there is change in entropy and volume. Therefore a first order phase transition can be defined as that one in which the Gibbs function with respect to pressure and temperature change discontinuously at the transition point. However, the value of the Gibbs function is the same in both the phases at equilibrium. These changes are represented graphically in Fig. 1.35.

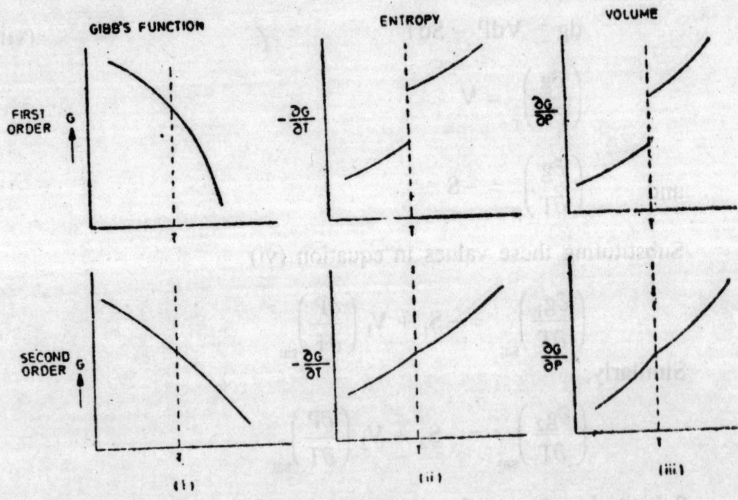

Fig. 1.35

1.67 THE T dS EQUATION

(1) The entropy S of a pure substance can be taken as a function of temperature and volume.

$$\therefore \qquad dS = \left(\frac{\partial S}{\partial T}\right)_V = dT + \left(\frac{\partial S}{\partial V}\right)_T dV \qquad \text{...(i)}$$

Multiplying both sides by T

$$T.dS = \left(\frac{\partial S}{\partial T}\right)_V dV + T\left(\frac{\partial S}{\partial V}\right)_T dV \qquad \text{...(ii)}$$

$$\therefore \text{ But } \qquad C_V = T\left(\frac{\partial S}{\partial T}\right)_V$$

and from Maxwell's relations

$$\left(\frac{\partial S}{\partial V}\right)_V = \left(\frac{\partial P}{\partial T}\right)_V$$

Substituting these values in equation (ii)

$$T\,dS = C_V dT + T\left(\frac{\partial P}{\partial T}\right)_V dV \qquad \text{...(iii)}$$

Equation (iii) is called the first TdS equation.

(2) The entropy S of a pure substance can also be regarded as a function of temperature and pressure

$$\therefore \qquad dS = \left(\frac{\partial S}{\partial T}\right)_P dT + \left(\frac{\partial S}{\partial P}\right)_T dP \qquad \text{...(iv)}$$

Multiplying both sides by T

$$T.dS = T\left(\frac{\partial S}{\partial T}\right)_P dT + T\left(\frac{\partial S}{\partial P}\right)_T dP \qquad \text{...(v)}$$

$$\text{But} \qquad C_P = T\left(\frac{\partial S}{\partial T}\right)_P$$

and from Maxwell's relations

$$\left(\frac{\partial S}{\partial P}\right)_T = -\left(\frac{\partial V}{\partial T}\right)_P$$

Substituting these values in equation (v)

$$T\,dS = C_P\,dT - T\left(\frac{\partial V}{\partial T}\right)_P \qquad ...(vi)$$

Equation (vi) is called the second TdS equation.

SOLVED EXAMPLES

Example 1:

When a system is taken from the state A to the state B, along the path ACB, 80 joules of heat flows into the system, and the system does 30 joules of work (Fig. 1.36).

(a) Bow much heat flows into the system along the path ADB, if the work done is 10 joules.

(b) The system is returned from the state B to the state A along the curved path. The work done on the system is 20 joules. Does the system absorb or liberate heat and how much?

(c) If $U_A = 0$, $U_D = 40$ joules, find the heat absorbed in the process AD and DB.

Solution:

Along the path ACB,

$HACB = U_B - U_A + W$

Here $H = +80$ joules

$W = +30$ joules

$\therefore +80 = U_B - U_A + 30$

$U_B - U_A = 80 - 30 = 50$ joules

(a) Along the path ADB,

$W = +10$ joules

$H_{ADB} = U_B - U_A + W$

$H = 50 + 10 = 60$ joules

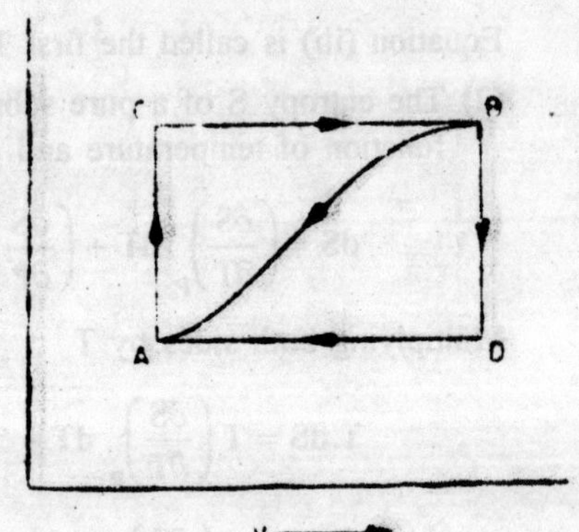

Fig. 1.36

(b) For the curved path from B to A,

$W = -20$ joules

$H = (U_A - U_B) + W$

$= -50 - 20 = -70$ joules

(–ve sign shows that heat is liberated by the system)

(c) $U_A = 0,$

$U_D = 40$ joules

$U_B - U_A = 50$

$\therefore$ $U_B = 50$ joules

In the process ADB, 10 joules of work is done. Work done from 4 to D is +10 joules and from D to B is zero.

For AD,

$H_{AD} = (U_D - U_A) + W$

$= 40 + 10 = 50$ joules

For DB

$H_{DB} = U_C - U_D + W$

$= \mathbf{50 - 40 + 0 = 10}$ **joules.**

Example 2:

A motor car tyre has a pressure of 2 atmospheres at the room temperature of 27°C. If the tyre suddenly bursts, find the resulting temperature.

Solution:

Here, $P_1 = 2$ atmospheres

$T_1 = 273 + 27$

$= 300$ K

$P_2 = 1$ atmosphere

$T_2 = ?$

$\gamma = 1.4$

$$\frac{P_1^{\gamma-1}}{T_1^{\gamma}} = \frac{P_2^{\gamma-1}}{T_2^{\gamma}}$$

$$\left(\frac{P_2}{P_1}\right)^{\gamma-1} = \left(\frac{T_2}{T_1}\right)^{\gamma}$$

$$\left(\frac{1}{2}\right)^{0.4} = \left(\frac{T_2}{300}\right)^{1.4}$$

$$0.4 \log (0.5) = 1.4 [\log T_2 - \log 300]$$

$$-0.1204 = 1.4 \log T_2 - 3.4680$$

$$1.4 \log T_2 = 3.4680 - 0.1204$$

$$= 3.3476$$

$$\log T_2 = \frac{3.3476}{1.4}$$

$$= 2.3911$$

$$T_2 = 246.1 \text{ K}$$

$$= \mathbf{-26.9°C.}$$

Example 3:

A quantity of air at 27°C and atmospheric Pressure it suddenly compressed to half its original volume, find the final (i) pressure and (ii) temperature.

Solution:

(i) $P_1 - 1$ atmosphere; P_2 =?, $\gamma = 1.4$

$V_1 = V$;

$V_2 = V/2$

During sudden compression, the process is adiabatic

$$P_1V_1^\gamma = P_2V_2^\gamma$$

$$P_2 = P_1\left[\frac{V_1}{V_2}\right]^\gamma$$

$$= 1[2]^{1/4}$$

$$= 2.636 \text{ atmospheres}$$

(ii) $V_1 = V$;

$V_2 = V/2$

$T_1 = 300$ K;

$$T_2 = ?$$

$$\gamma = 1.4$$

$$T_1(V_1)^{\gamma-1} = T_2\,(V_2)^{\gamma-1}$$

$$T_2 = T_1[2]^{1/4-1}$$

$$= 300[2]^{0.4}$$

$$= 395.9 \text{ K}$$

$$= \mathbf{122.9°C.}$$

Example 4:

Air is compressed adiabatically to half its volume. Calculate the change in its temperature.

Solution:

Let the initial temperature be T_1 K and the final temperature

Initial volume $= V_1$

Final volume $= V_2$

$= V_1/2$

During an adiabatic process

$$T_1V_1^{\gamma-1} = T_2V_2^{\gamma-1}$$

$$T_2 = T_1\left[\frac{V_1}{V_2}\right]^{\gamma-1}$$

$$T_2 = T_1[2]^{\gamma-1}$$

But γ for air = 1.40

$$T_2 = T_1[2]^{1.40-1}$$

$$T_2 = T_1[2]^{0.40}$$

$$T_2 = 1.319\,T_1$$

Change in temperature

$$= T_2 - T_1$$

$$= 1.319\,T_1 - T_1$$

$$= \mathbf{1.319\,T_1\ K.}$$

Example 5:

1 gram molecule of a monoatomic (γ = 5/3) perfect gas at 27°C is adiabatically compressed in a reversible process from an initial pressure of 1 atmosphere to a final pressure of 60 atmospheres. Calculate the resulting difference in temperature.

Solution:

In a reversible adiabatic process

$$\frac{P_1^{\gamma-1}}{T_1^{\gamma}} = \frac{P_2^{\gamma-1}}{T_2^{\gamma}}$$

or $$\left(\frac{P_2}{P_1}\right)^{\gamma-1} = \left(\frac{T_2}{T_1}\right)^{\gamma}$$

Here, $P_2 = 50,$

$P_1 = 1,$

$T_1 = 273 + 27$

$= 300$ K

$T_2 = ?$

$\gamma = 5/3$

$$\therefore \quad (50)^{2/3} = \left(\frac{T_2}{300}\right)^{5/3}$$

$$2/3 \log (50) = 5/3 \, [\log T_2 - \log 300]$$

$$T_2 = 1{,}434 \text{ K}$$

$$= 1{,}161°\text{C}.$$

Example 6:

A quantity, of dry air at 27°C is compressed (i) slowly and (ii) suddenly to 1/3 of its volume, find the change in temperature in each case, assuming f to be 1.4 for dry air.

Solution:

(1) When the process is slow, the temperature of the system remains constant. Therefore, there is no change in temperature.

(2) When the compression is sudden, the process is adiabatic.

Here $V_1 = V,$

$V_2 = V/3$

$T_1 = 300$ K,

$T_2 = ?$

$\gamma = 1.4$

$$T_2(V_2)^{\gamma-1} = T_1[V_1]^{\gamma-1}$$

$$T_2 = T_1\left[\frac{V_1}{V_2}\right]^{\gamma-1}$$

or $$T_2 = 300\left[\frac{3V}{V}\right]^{\gamma-1}$$

$$= 300\ [3]1.4.1$$

or $$T_2 = 465.5 \text{ K}$$

$$= 192.5°\text{C}$$

The temperature of air increases by

192.5–27 = 165.5°C or 165.5 K.

Example 7:

A certain mass of gas at NTP is expanded to three times its volume under adiabatic conditions. Calculate the resulting temperature and pressure, γ for the gas is 1.40.

Solution:

(1) Here, $V_1 = V,$

$V_2 = 3V$

$T_1 = 273$ K

$T_2 = ?$

$$T_1\ V_1^{\gamma-1} = T_2\ V_2^{\gamma-1}$$

or $$T_2 = T_1\left[\frac{V_1}{V_2}\right]^{\gamma-1}$$

$$T_2 = 273\left[\frac{1}{3}\right]^{1.4-1}$$

$$\mathbf{T = 276\ K = -97°C.}$$

(2) Here, $V_1 = V$,

$$V_2 = 3V$$

$$P_1 = 1 \text{ atmosphere,}$$

$$P_2 = ?$$

$$P_1\, V_1^{\gamma} = P_2\, V_2^{\gamma}$$

$$P_2 = P_1\left[\frac{V_1}{V_2}\right]^{\gamma}$$

or $$P_2 = 1\left(\frac{1}{3}\right)^{1.4}$$

$$\mathbf{P_2 = 0.2148\ atmosphere.}$$

Example 8:

Find the efficiency of the carnot's engine working between the steam point and the ice point.

Solution:

We have $T_1 = 273 + 100 = 373\ K$

$$T_2 = 273 + 0 \quad = 273\ K$$

$$\eta = 1 - \frac{T_2}{T_1}$$

$$= 1 - \frac{273}{373} = \frac{100}{373}$$

$$\% \text{ efficiency} \quad = \frac{100}{373} \times 100$$

$$\mathbf{= 26.81\%.}$$

Example 9:

Find the efficiency of a carnot's engine working between 127°C and 27°C.

Solution:

We have $T_1 = 273 + 127 = 400$ K

$T_2 = 273 + 27 = 300$ K

$$\eta = 1 - \frac{T_2}{T_1}$$

$$= 1 - \frac{300}{400} = 0.25$$

% efficiency = **25%.**

Example 10:

A carnot's engine whose temperature of the source is 400 K takes 200 calories of heat at this temperature and rejects 150 calories of heat to the sink. What is the temperature of the sink? Also calculate the efficiency of the engine.

Solution:

We have $H_1 = 200$ cal;

$H_2 = 150$ cal

$T_1 = 400$ cal;

$T_2 = ?$

$$\frac{H_1}{T_1} = \frac{H_2}{T_2}$$

$$T_2 = T_2 \frac{H_2}{H_1} \times 400 = 300 \text{ K}$$

$$\eta = 1 - \frac{T_2}{T_1}$$

$$= 1 - \frac{300}{400} = 0.25$$

% efficiency = **25%.**

Example 11:

A Carnot's engine is operated between two reservoirs at temperatures of 450 K and 350 K. If the engine receives 1000 calories of heat from the source in each cycle, calculate the amount of heat rejected to the

sink in each cycle. Calculate the efficiency of the engine and the work done by the engine in each cycle. (1 calorie = 4.2 joules).

Solution:

We have $T_1 = 450$ K;

$T_2 = 350$ K

$H_1 = 1000$ cal;

$H_2 = ?$

$$\frac{H_2}{H_1} = \frac{T_2}{T_1}$$

$$H_2 = H_1 \times \frac{T_2}{T_1}$$

$$= \frac{1000 \times 350}{450} = \mathbf{777.77\ cals}$$

$$\eta = 1 - \frac{T_2}{T_1}$$

$$= 1 - \frac{350}{450} = \frac{100}{450}$$

$$= 0.2222$$

% efficiency $\quad$ = **22.22%.**

Work done in each cycle

$= H_1 - H_2$

$= 1000 - 777.77$

$= 222.23$ cal

$= 222.23 \times 4.2$ joules

= 933.33 Joules.

Example 12:

A carnot's engine working as a refrigerator between 260 k and 300 k receives 500 calories of heat from the reservoir at the lower temperature. Calculate the amount of heat rejected to the reservoir at the higher temperature. Calculate also the amount of work done in each cycle to operate the refrigerator.

Solution:

We have $H_1 = ?$

$H_2 = 500$ cal

$T_1 = 300$ K

$T_2 = 260$ K

$$\frac{H_1}{H_2} = \frac{T_1}{T_2};$$

$$H_1 = H_2 \cdot \frac{T_1}{T_2}$$

$$H_1 = \frac{500 \times 300}{260} = 576.92 \text{ cal}$$

$$W = H_1 - H_2 = 76.92 \text{ cal}$$

$$= 76.92 \times 4.2 \text{ joules}$$

$$= \mathbf{323.08 \text{ joules.}}$$

Example 13:

A carnot's refrigerator takes heat from water at 0°C and discards it to a room at 27°C. 1 kg of water at 0°C is to be changed into ice at 0°C. How many calories of heat are discarded to the room? What is the work done by the refrigerator in this process? What is the coefficient of performance of the machine?

Solution:

We have $H_1 = ?$

$H_2 = 1000 \times 80 = 80{,}000$ cal

$T_1 = 300$ K

$T_2 = 273$ K

(1) $$\frac{H_1}{H_2} = \frac{T_1}{T_2}$$

$$H_1 = \frac{H_2 T_1}{T_2}$$

$$= \frac{80,000 \times 300}{273}$$

$$\mathbf{H_1 = 87,900\ cal.}$$

(2) Work done by the refrigerator

$$= W = J\,(H_1 - H_2)$$

$$W = 4.2\,(87,900 - 80,000)$$

$$W = 4.2 \times 7900$$

or $$W = 3.183 \times 10^4 \text{ joules}$$

(3) Coefficient of performance,

$$= \frac{H_2}{H_1 - H_2}$$

$$= \frac{80,000}{87,900 - 80,000}$$

$$= \frac{80,000}{7900}$$

$$= \mathbf{10.13.}$$

Example 14:

A carnot engine whose low temperature reservoir is at 7°C has an efficiency of 50%. It is desired to increase the efficiency to 70%. By how many degrees should the temperature of the high temperature reservoir be increased?

Solution:

In the first case

$$\eta = 50\% = 0.5,$$

$$T_2 = 273 + 7 = 280 \text{ K}$$

$$T_1 = ?$$

$$\eta = 1 - \frac{T_2}{T_1}$$

or $$0.5 = 1 - \frac{280}{T_1}$$

or $$T_1 = 560 \text{ K}$$

In the second case

$$\eta' = 70\% = 0.7$$

$$T_2 = 280 \text{ K},$$

$$T_1' = ?$$

$$\eta' = 1 - \frac{T_2}{T_1'}$$

$$0.7 = 1 - \frac{280}{T_1'}$$

or $$T_1' = 840 \text{ K}$$

Increase in temperature = 840 – 560 = **280 K.**

Example 15:

An inventor claims to have developed an engine working between 600 K and 300 K capable of having an efficiency of 52%, Comment on his claim.

Solution:

The efficiency of a Carnot's engine working between 600 K and 300 K,

$$\eta = 1 - \frac{T_2}{T_1}$$

$$= 1 - \frac{300}{600} = 0.5$$

$$= 50\%$$

The efficiency claimed = η = 52%.

It means that the efficiency of the engine is more than the efficiency of a Carnot's engine working between the same two temperature limits. But, no engine can have an efficiency more than a Carnot's engine, so his claim is invalid.

Example 16:

In a double acting steam engine, the average pressure of steam is 10^5 newtons/metre^2. The length of the stroke if 1 metre and the area of the piston is 0.15 sq metre. Find the power of the engine, if it makes 5 strokes per second.

Solution:

Here $P = W$ newtons/metre^2

$L = 1$ metre

$A = 0.15$ sq metre

$N = 5$ strokes/second

$$\text{Power} = \frac{2\,PLAN}{1000}$$

$$\text{Power of the engine} = \frac{2 \times 10^5 \times 1 \times 0.15 \times 5}{1000}$$

$$= \textbf{150 kilowatts.}$$

Example 17:

Calculate the depression in the melting point of ice produced by one atmosphere increase of pressure. Given latent heat of ice = 80 cal per gram and the specific volumes of 1 gram of ice and water at 0°C are 1.091 cm^3 and 1.000 cm^3 respectively.

Solution:

Here $L = 80 \text{ cal} = 80 \times 4.2 \times 10^7$ ergs

$T = 273$ K

$dP = 1$ atmosphere

$= 76 \times 13.6 \times 980$ dynes/cm^2

$V_1 = 1.091$ cm^3

$V_2 = 1.000$ cm^3

$$\frac{dP}{dT} = \frac{L}{T\,(V_2 - V_1)}$$

$$dT = \frac{dP.T.(V_2 - V_1)}{L}$$

$$= \frac{76 \times 13.6 \times 980 \times 273\,(1 - 1.091)}{80 \times 4.2 \times 10^7}$$

$$= 0.0074 \text{ K.}$$

Therefore, the decrease in the melting point of ice with an increase in pressure of one atmosphere

$$= \mathbf{0.0074\ K = 0.0074°C.}$$

Example 18:

Find the increase in the boiling point of water at 100°C when the pressure is increased by one atmosphere. Latent heat of vaporisation of steam is 640 cal/gram and I gram of steam occupies a volume of 1677 cm³.

Solution:

We have $dP = 76 \times 13.6 \times 980$ dynes/cm²

$$T = 100 + 273$$

$$= 373 \text{ K}$$

$$L = 540 \times 4.2 \times 10^7 \text{ ergs}$$

$$V_1 = 1.000 \text{ cm}^3$$

$$V_2 = 1677 \text{ cm}^3$$

$$\frac{dP}{dT} = \frac{L}{T(V_2 - V_1)}$$

$$dT = \frac{dP \times T(V_2 - V_1)}{L}$$

$$= \frac{76 \times 13.6 \times 980 \times 373 \times 1676}{540 \times 4.2 \times 10^7}$$

$$= \mathbf{27.92°C.}$$

Therefore, the increase in the boiling point of water with an increase in pressure of one atmosphere

$$= \mathbf{27.92°C}$$

$$= \mathbf{27.92\ K.}$$

Example 19:

Calculate the change in temperature of boiling water when the pressure is increased by 27.12 mm of Hg. The normal boiling point of water at atmospheric pressure is 100°C.

Latent heat of steam = 537 cal/g

and specific volume of steam = 1674 cm³.

Solution:

We have $dP = 2.712 \times 13.6 \times 980$ dynes/cm²

$$T = 100 + 273 = 373 \text{ K}$$

$$L = 537 \times 4.2 \times 10^7 \text{ ergs}$$

$$V_1 = 1.000 \text{ cm}^3$$

$$V_2 = 1674 \text{ cm}^3$$

$$\frac{dP}{dT} = \frac{L}{T(V_2 - V_1)}$$

$$dT = \frac{dP \times T\,(V_2 - V_1)}{L}$$

$$= \frac{2.712 \times 13.6 \times 980 \times 373 \times 1673}{537 \times 4.2 \times 10^7}$$

= 1°C or 1 K.

Example 20:

Calculate the change in the melting point of naphthalene for one atmosphere rise in pressure given that its melting point is 80°C. Latent heat of fusion is 4563 cal/mol and increase in volume on fusion is 18.7 cm³ per mol and 1 calorie = 4.2 × 10⁷ ergs.

Solution:

We have $\frac{dP}{dT} = \frac{L}{T(V_2 - V_1)}$

$$dT = 1 \text{ atmosphere}$$

$$= 76 \times 13.6 \times 980 \text{ dynes/cm}^2$$

$dT = ?$

$L = 4563$ cal/mol

$= 4563 \times 4.2 \times 10^7$ ergs/mol

$V_2 - V_1 = 18.7$ cm^3/mol

$T = 80 + 273$

$= 353$ K

$$dT = \frac{dP.T.(V_2 - V_1)}{L}$$

$$= \frac{76 \times 13.6 \times 980 \times 373 \times 18.7}{4563 \times 4.2 \times 10^7}$$

= 0.03488 K.

Therefore, the increase in the melting point of naphthalene with an increase in pressure of one atmosphere

= 0.03488 K

= 0.03488°C.

Example 21:

Calculate the boiling point of benzene under a pressure of 80 cm of mercury. The normal boiling point is 80°C. Latent heat of vaporization is 389 joules/g, density of vapour at the boiling point is 4 g/litre and that of the liquid 0.9 g/cm³.

Solution:

Here $dP = 80 - 76$

$= 4$ cm of Hg

$= 4 \times 13.6 \times 980$ dynes/cm^2

$T = 80°C = 80 + 273$

$= 353$ K

$E = 380$ joules/g

$= 380 \times 10^7$ ergs/g

$$V_1 = \frac{1}{0.9} = 1.11 \text{ cm}^3$$

$$V_2 = \frac{1,000}{4} = 250 \text{ cm}^3$$

$$\frac{dP}{dT} = \frac{L}{T[V_2 - V_1]}$$

$$dT = \frac{dP \times T[V_2 - V_1]}{L}$$

$$dT = \frac{4 \times 13.6 \times 980 \times 373 \times [250 - 1.11]}{380 \times 10^7}$$

= 1.233 K

= 1.233°C.

∴ The boiling point of benzene at a pressure of 80 cm of Hg

= 80 + 1.233

= 81.233°C.

Example 22:

Calculate the change in the boiling point of water when the pressure of steam on its surface is increased from 1 atmosphere to 1.10 atmospheres.

Latent heat of water at 100°C

= 537 cal/g

Volume of one-gram of steam at 100°C

= 1.676 cm³.

Solution:

We have dP = 1.10 − 1

= 0.10 atmosphere

$dP = 0.1 \times 76 \times 13.6 \times 980$ dynes/cm²

T = 100 + 273

= 373 K

L = 537 cals/g

$= 537 \times 4.2 \times 10^7$ ergs/g

$V_1 = 1.00$ cm³

$$V_1 = 1.676 \text{ cm}^3$$

$$\frac{dP}{dT} = \frac{L}{T[V_2 - V_1]}$$

$$dT = \frac{dP \times T[V_2 - V_1]}{L}$$

$$dT = \frac{0.1 \times 76 \times 13.6 \times 980 \times 373 \times (1{,}675)}{537 \times 4.2 \times 10^7}$$

$$= \mathbf{2.792\ K}$$

$$= \mathbf{2.792°C.}$$

Therefore, the increase in the boiling point of water with an increase of 0.1 atmosphere pressure

$$= \mathbf{2.792\ K}$$

$$= \mathbf{2.792°C.}$$

Example 23:

Calculate the change in the melting point of ice when it is subjected to a pressure of 100 atmospheres.

Density of ice = *0.917 g/cm³, and*

Latent heat of ice = *336 J/g.*

Solution:

We have $\frac{dP}{dT} = \frac{L}{T(V_2 - V_1)}$

$$dP = 100 - 1$$

$$= 99 \text{ atmospheres}$$

$$dP = 99 \times 76 \times 13.6 \times 980 \text{ dynes/cm}^2$$

$$L = 336 \text{ J/g}$$

$$= 336 \times 10^7 \text{ ergs/g}$$

$$T = 273\text{K}$$

$$(V_2 - V_1) = 1 - \frac{1}{0.917}$$

$$= -\frac{0.083}{0.917}$$

$$= -0.091 \text{ cm}^3$$

$$\therefore \quad dT = \frac{TdP(V_2 - V_1)}{L}$$

$$= \frac{273 \times 99 \times 76 \times 13.6 \times 980 \times (-0.091)}{336 \times 10^7}$$

$$dT = -0.7326 \text{ K}$$

$$= \mathbf{-0.7326°C.}$$

The decrease in the melting point of ice with a pressure of 100 atmospheres

Example 24:

Calculate the pressure required to lowe melting point of ice by 0°C. (L = 79.6 cal/g, specific volume of water at 0°C = 1.000 cm² specific volume of ice at 0°C = 1.091 cm³ and 1 atmosphere pressure = 1.013 × 10⁶ dynes/cm²).

Solution:

$$\text{We have } \frac{dP}{dT} = \frac{L}{T(V_2 - V_1)}$$

$$dT = -1 \text{ K}$$

$$T = 273 \text{ K}$$

$$V_2 - V_1 = -0.091 \text{ cm}^3$$

$$L = 79.\text{-}6 \text{ cal/g}$$

$$= 79.6 \times 4.18 \times 10^7 \text{ ergs/g}$$

$$dP = \frac{L.dT}{T(V_2 - V_1)}$$

$$dP = \frac{79.6 \times 4.18 \times 10^7 \times 1}{273 \times 0.091} \text{ dynes/cm}^2$$

$$\text{or} \quad dP = \frac{79.6 \times 4.18 \times 10^7}{273 \times 0.091 \times 1013 \times 10^6} \text{ atmospheres}$$

$$dP = 135.2 \text{ atmospheres}$$

$\therefore$ Pressure required

$$= 135.2 + 1$$

$$= \mathbf{136.2 \text{ atmospheres.}}$$

Example 25:

Water boils at a temperature of 101°C at a pressure of 787 mm of Hg. 1 gram of water occupies 1,601 cm³ on evaporation. Calculate the latent heat of steam. J = 4.2 × 10⁷ ergs/cal.

Solution:

We have $\dfrac{dP}{dT} = \dfrac{L}{T(V_2 - V_1)}$

$$dP = 787 - 760$$

$$= 27 \text{ mm of Hg}$$

$$= 2.7 \text{ cm of Hg}$$

$$= 2.7 \times 13.6 \times 980 \text{ dynes/cm}^2$$

$$dT = 1°C = 1K$$

$$T = 373 \text{ K}$$

$$V_2 - V_1 = 1{,}601 - 1 = 1{,}600 \text{ cm}^3$$

$$L = ?$$

$$L = \frac{L\,dP\,(V_2 - V_1)}{dT}$$

$$L = \frac{373 \times 2.7 \times 13.6 \times 980 \times 1{,}600}{1} \text{ ergs/g}$$

or $$L = \frac{373 \times 2.7 \times 13.6 \times 980 \times 1{,}600}{4.2 \times 10^7} \text{ cal/g}$$

$$\mathbf{L = 511.3 \text{ cal/g.}}$$

Example 26:

When lead is melted at atmospheric pressure, (the melting point is 600 K) the density decreases from 11.01 to 10.66 g/cm³ and the latent heat of fusion is 24.5 J/g. What is the melting point at a pressure of 100 atmospheres?

Solution:

We have $\frac{dP}{dT} = \frac{L}{T(V_2 - V_1)}$

$$dP = 99 \text{ atmospheres}$$

$$= 99 \times 76 \times 13.6 \times 980 \text{ dynes/cm}^2$$

$$L = 24.5 \text{ J/g}$$

$$= 24.5 \times 10^7 \text{ ergs/g}$$

$$V_1 = \frac{1}{11.01} \text{ cm}^3$$

$$V_2 = \frac{1}{10.65} \text{ cm}^3$$

$$V_2 - V_1 = \frac{1}{10.65} - \frac{1}{11.01}$$

$$= \left(\frac{0.36}{10.65 \times 10.01}\right) \text{cm}^3$$

$$dT = \frac{TdP\,(V_2 - V_1)}{L}$$

$$dT = \frac{600 \times 99 \times 76 \times 13.6 \times 980 \times 0.36}{10.65 \times 11.01 \times 24.5 \times 10^7}$$

$$\mathbf{dT = 0.7539\ K}$$

$$\mathbf{= 0.7539°C.}$$

∴ Melting point of lead at 100 atmospheres pressure

$$= 600 + 0.7539$$

$$\mathbf{= 600.7539\ K.}$$

Example 27:

Calculate under what pressure water will boil at 120°C, if the change in specific volume when 1 gram of water is converted into steam is 1,676 cm³.

Latent heat of steam

= 540 cal/g

J = 4.2 × 10^7 ergs/cal

1 atmosphere pressure

= 10^6 dynes/cm^2.

Solution:

Here $$\frac{dP}{dT} = \frac{L}{T(V_2 - V_1)}$$

$$dT = 120 = 100$$

$$= 20 \text{ K}$$

$$T = 373 \text{ K}$$

$$(V_2 - V_1) = 1{,}676 \text{ cm}^3$$

$$L = 540 \text{ cal/g}$$

$$= 540 \times 4.2 \times 10^7 \text{ ergs/g}$$

$$dP = ?$$

$$dP = \frac{L.dT}{T(V_2 - V_1)}$$

$$dP = \frac{540 \times 4.2 \times 10^7 \times 20}{373 \times 1{,}676} \text{ dynes/cm}^2$$

$$dP = \frac{540 \times 4.2 \times 10^7 \times 20}{373 \times 1{,}676 \times 10^6} \text{ atmospheres}$$

$$dP = 0.7254 \text{ atmosphere}$$

Pressure required = 1 + 0.7254

= 1 + 0.7254 atmospheres.

Example 28:

Calculate the change in entropy when 10 grams of ice at 0°C is converted into water at the same temperature.

Solution:

Heat absorbed by 10 g of ice at 0°C when it is converted into water at 0°C = 10 × 80 = 800 cal

∴ $\delta H = 800$ cal

$T = 0°C = 273$ K

The gain in entropy

$$dS = \frac{\delta H}{T}$$

$$= \frac{800}{273} = 2.93 \text{ cal/K}.$$

Example 29:

Calculate the change in entropy when 5 kg of water at 100°C is converted into steam at the same temperature.

Solution:

Heat absorbed by 5 kg of water at 100°C when it is converted into steam at 100°C

$= 5000 \times 540$

$= 2700000$ cal

SB = 2700000 cal

The gain in entropy

$$dS = \frac{\delta H}{T}$$

$$= \frac{270000}{373} = 72 \text{ cal/K.}$$

Example 30:

Calculate the increase in entropy when 1 gram of ice at –10°C is Converted into steam at 100°C. Specific heat of ice a 0.5, latent heat of ice = 80 cal/g, latent heat of steam = 540 cal/g.

Solution:

(1) Increase in entropy when the temperature of 1 gram of ice increases from –10°C to 0°C

$$dS = \int_{T_1}^{T_2} \frac{\delta H}{T}$$

$$= ms\int_{T_1}^{T_2} \frac{dT}{T}$$

$$= ms \log_e \frac{T_2}{T_1}$$

$$= ms \times 2.3026 \log_e \frac{T_2}{T_1}$$

$$= 1.0.5 \times 2.3026 \log_{10} \frac{273}{263}$$

$$\mathbf{= 0.01865\ cal/K.}$$

(2) Increase in entropy when 1 gram of ice at 0°C is converted into water at 0°C.

$$dS = \frac{\delta H}{T}$$

$$= \frac{80}{273} = 0.293 \text{ cal/K}$$

(3) Increase in entropy when the temperature of 1 f of water is raised from 0°C to 100°C.

$$dS = \int_{T_1}^{T_2} \frac{\delta H}{T}$$

$$= ms \times 2.3026 \log_{10} T_2/T_1$$

$$= 1 \times 1 \times 2.3026 \log_{10} 373/273$$

$$\mathbf{= 0.312\ cal/K.}$$

(4) Increase in entropy when 1 g water at 100°C is converted into steam at 100°C

$$dS = \frac{\delta H}{T}$$

$$= \frac{540}{373} = 1.447 \text{ cal/K}$$

Total increase in entropy

$$= 0.01865 + 0.293 + 0.312 + 1.447$$

$$\mathbf{= 2.07065\ cal/K.}$$

Example 31:

One gram molecule of a gas expands isothermally to four times its volume. Calculate the change in its entropy in terns of the gas constant.

Solution:

Work done $= \int_{V_1}^{V_2} PdV$

But $PV = RT$

or $P = \frac{RT}{V}$

$$W = RT \int_{V_1}^{V_2} \frac{dV}{V}$$

$$= RT \log_e \frac{V_2}{V_1}$$

Here $\frac{V_2}{V_1} = 4$

$$W = RT \times 2.3026 \log_{10}(4)$$

Here, W and R are in the units of work

Gain in entropy = δH/T

$$= \frac{W}{JT} = \frac{RT \times 2.3026 \log_{10} 4}{JT}$$

$$\mathbf{= 1.387\ R/J\ cal/K.}$$

Example 32:

50 grams of water at 0°C if mixed with an equal mass of water at 83°C. Calculate the resultant increase in entropy.

Solution:

(i)$m_1 = 50$ g; $T_1 = 273$ K

$m_1 = 50$ g; T, = 353 K

Let the final temperature of the mixture be T K

$$m_1\ 8 \times (T - T_1) = m_2\ s(T_2 - T)$$

$$50 \times 1 \times (T - 273) = 50 \times 1 \times (353 - T)$$

$$\mathbf{T = 310\ K.}$$

(ii) Change in entropy by 50 g of water when its temperature from 273 K to 313 K.

$$= \frac{\delta H}{T}$$

$$= ms \int_{T_1}^{T} \frac{dT}{T}$$

$$= 50 \times 1 \times \log_e \frac{313}{273}$$

$$= 50 \times 2.3026 \times \log_e \frac{313}{273}$$

$$= \mathbf{+6.829\ cal/K.}$$

Here, the +ve sign indicates gain in entropy.

(iii) Change in entropy by 50 g of water when its temperature falls from 353 K to 313 K

$$= \frac{\delta H}{T} = ms \int_{T_2}^{T} \frac{dT}{T}$$

$$= 50 \times 1 \times \log_e \frac{313}{353}$$

$$= 50 \times 2.3026 \times \log_{10} \frac{313}{353}$$

$$= -6.023\ cal/K.$$

Here, the –ve sign indicates loss in entropy.

Therefore, the total gain in entropy of the system

$$= 6.829 - 6.023$$

$$= \mathbf{0\ 806\ cal/K.}$$

Example 33:

Calculate the change in entropy when 50 grams of water at 15°C is mixed with 80 grams of water at 40°C. Specific heat of wetter may be assumed to be equal to 2.

Solution:

(i) $m_1 = 50g$

$T_1 = 15 + 273 = 288$ K

$m_2 = 10$ grams

$T_2 = 40 + 273 = 313$ K.

Let the final temperature be T K.

$$m_1 \times 8 \times (T - T_1) = m_2 \times 8 \times (T_2 - T)$$

$$50 \times 1 \times (T - 288) = 80 \times 1 \times (313 - T)$$

$$\mathbf{T = 303\text{-}4\ K.}$$

(ii) Change in entropy when the temperature of 50 g of water rises from 288 K to 303.4 K

$$= \frac{\delta H}{T} = ms \int_{T_1}^{T} \frac{dT}{T}$$

$$= 50 \times 1 \times 2.3026 \times \log_{10} \frac{303.4}{288}$$

$$\mathbf{= +2.602\ cal/K.}$$

(iii) Change in entropy when the temperature of 80 g of water decreases from 313 K to 303.4 K

$$= \frac{\delta H}{T} = ms \int_{T_2}^{T} \frac{dT}{T}$$

$$= 80 \times 1 \times 2.3026 \times \log_{10} \frac{303.4}{313}$$

$$\mathbf{= -2.487\ cal/K.}$$

Therefore, the net change in the entropy of the system

$$= +2.602 - 2.487$$

$$= +0.115\ cal/K$$

Hence the net increase in the entropy of the system

$$\mathbf{= 0.115\ cal/K.}$$

Example 34:

10 g of steam at 100°C is blown into 90 grams of water at 0°C, contained in a calorimeter of water equivalent 19 grams. The whole of the steam is condensed. Calculate the increase in the entropy of the system.

Solution:

(i) $m_1 = 10g$

$T_1 = 100°C = 373\ K$

$m_2 = 90 + 10 = 100g$

$T_2 = 273$ K

Let the final temperature be T K

$10 \times 540 + 10(373 - 7) = 100(T - 273)$

$$T = 331.2 \text{ K.}$$

(ii) Change in entropy when the temperature of water and calorimeter rises from 273 K to 331.2 K

$$= \frac{\delta H}{T} = ms \int_{T_2}^{T} \frac{dT}{T}$$

$$= 100 \int_{273}^{331.2} \frac{dT}{T}$$

$$= 100 \times 2.3026 \times \log_{10}\left(\frac{331.2}{273}\right)$$

$$= +19.32 \text{ cal/K.}$$

(iii) Change in entropy when 10 grams of steam at 373 K is condensed to water at 373 K

$$= \left(\frac{\delta H}{T}\right) = -\frac{10 \times 540}{273}$$

$$= -14.47 \text{cal/K.}$$

(–ve sign indicates decrease in entropy).

(iv) Change in entropy when 10 grams of water at 373 K is cooled to water at 331.2 K

$$= \frac{\delta H}{T} = ms \int_{T_2}^{T} \frac{dT}{T}$$

$$= 10 \times 2.3026 \log_{10}\left(\frac{331.2}{373}\right)$$

$$= -1.188 \text{ cals/K}$$

Net change in entropy

$$= 19.32 - 14.47 - 1.188$$

$$= +3.662 \text{ cal/K.}$$

Hence the net increase in the entropy of the system

$$= 3.662 \text{ cal/K.}$$

Example 35:

1 g of water at 20°C is converted into ice at –10°C at constant pressure. Heat capacity for 1 g of water is 4.2 J/g-K and that of ice is 2.1 J/g-K. Beat of fusion of ice at 0°C = 335 J/g. Calculate the total change in the entropy of the system.

Solution:

(i) Change in entropy when the temperature of 1 g of water at 293 K falls to 273 K.

$$dS = \frac{\delta H}{T} = ms \int_{T_1}^{T_2} \frac{dT}{T}$$

$$= 1 \times 4.2 \int_{293}^{273} \frac{dT}{T}$$

$$= 4.2 \times 2.3026 \log_{10} \left(\frac{273}{293}\right)$$

$$= \mathbf{-0.2969\ J/K.}$$

(ii) Change in entropy when 1 g of water at 273 K is converted into ice at 273 K

$$dS = \frac{\delta H}{T} = \frac{-1 \times 335}{273} = -1.227\ J/K$$

(iii) Change in entropy when the temperature of 1 g of ice at 273 K falls to 263 K

$$dS = \frac{\delta H}{T} = ms \int_{T_1}^{T_2} \frac{dT}{T}$$

$$= 1 \times 2.1 \times 2.3026 \log_{10} \left(\frac{263}{273}\right)$$

$$= -0.07834\ J/K$$

Total change in entropy of the system

$$= -0.2969 - 1.227 - 0.07834$$

$$= \mathbf{-1.60224\ J/K.}$$

Negative sign shows that there is decrease in entropy of the system.

Example 36:

1 kg of wafer of 273 K is brought in contact with a heat reservoir at 873 K:

(1) what is the change in entropy of water when its temperature reaches 873 K?

(2) What is the change in entropy of (i) the reservoir and (ii) the universe.

Solution:

(1) Increase in entropy when the temperature of 1000 g of water is raised from 273 K to 373 K

$$dS = \int_{T_1}^{T_2} \frac{\delta H}{T}$$

$$= ms \times 2.3026 \log_{10} T_2/T_1$$

$$= 1000 \times 1 \times 2.3026 \log_{10} 373/273$$

$$= \mathbf{312\ cal/K.}$$

(2) (i) Change in entropy of the reservoir,

$$dS = \frac{-\delta H}{T}$$

$$= -\frac{1000 \times 1 \times 100}{372} = \mathbf{-268.1\ cal/K.}$$

Negative sign shows decrease in entropy

(2) (ii) Change in entropy of the universe

$$= 312 - 268.1$$

$$= \mathbf{43.9\ cal/K.}$$

Therefore, the net increase in entropy of the universe

$$= \mathbf{43.9\ cal/K.}$$

Example 37:

Deduce Clapeyron's latent heat equation from Maxwell's thermodynamical relations.

Solution:

We have $\left(\frac{\partial H}{\partial V}\right)_T = T\left(\frac{\partial P}{\partial T}\right)_V$ ...(i)

Here $\left(\frac{\partial H}{\partial V}\right)_T$ represents the quantity of heat absorbed per unit increase in volume at constant temperature. This quantity of heat absorbed at constant temperature is the latent heat. If ∂H is the latent heat (L), when a unit mass of the substance changes in volume for V_1 to V_2 at constant temperature, then

$$\partial H = L \quad \text{and} \quad \partial V = V_2 - V_1$$

Substituting these values, in equation (i)

$$\therefore \quad \left(\frac{L}{V_2 - V_1}\right)_T = T\left(\frac{\partial P}{\partial T}\right)_V$$

or
$$\frac{L}{V_2 - V_1} = T\frac{dP}{dT}$$

or
$$\frac{dP}{dT} = \frac{L}{T(V_2 - V_1)} \qquad \text{...(ii)}$$

This is Clapeyron's latent heat equation.

Example 38:

Show that for a homogeneous fluid

$$C_P - C_V = T\left(\frac{\partial P}{\partial T}\right)_V\left(\frac{\partial V}{\partial T}\right)_P$$

Also show that for a perfect gas $C_P - C_V = R$ and for a gas obeying Van der Waah equation

Solution:

We have
$$C_P - C_V = R\left(1 + \frac{2a}{RTV}\right)$$

Here,
$$C_P\left(\frac{\partial H}{\partial T}\right)_P = T\left(\frac{\partial S}{\partial T}\right)_P$$

and
$$C_V\left(\frac{\partial H}{\partial T}\right)_V = T\left(\frac{\partial S}{\partial T}\right)_V$$

Considering S as a function of temperature and volume

$$dS = \left(\frac{\partial S}{\partial T}\right)_V = dT\left(\frac{\partial S}{\partial T}\right)_T dV$$

$$\left(\frac{\partial S}{\partial T}\right)_P = \left(\frac{\partial S}{\partial T}\right)_V \left(\frac{\partial T}{\partial T}\right) + \left(\frac{\partial S}{\partial V}\right)_T \left(\frac{\partial V}{\partial T}\right)_P$$

But from equation (v) [§ 6.52] $\left(\frac{\partial S}{\partial V}\right)_T = \left(\frac{\partial P}{\partial T}\right)_V$

$$T\left(\frac{\partial S}{\partial T}\right)_P = T\left(\frac{\partial S}{\partial T}\right)_V + T\left(\frac{\partial P}{\partial T}\right)_V \left(\frac{\partial V}{\partial T}\right)_P$$

$$C_P = C_V + T\left(\frac{\partial P}{\partial T}\right)_V \left(\frac{\partial V}{\partial T}\right)_P$$

$$\therefore \quad C_P - C_V = T\left(\frac{\partial P}{\partial T}\right)_V \left(\frac{\partial V}{\partial T}\right)_P \qquad \text{...(i)}$$

For a Perfect Gas

$$PV = RT$$

$$\therefore \quad \left(\frac{\partial P}{\partial T}\right)_V = \frac{R}{V} \text{ and } \left(\frac{\partial V}{\partial T}\right)_P = \frac{R}{P}$$

But, $\quad C_P - C_V = T\left(\frac{\partial P}{\partial T}\right)_V \left(\frac{\partial V}{\partial T}\right)_P$

$$C_P - C_V = T\left(\frac{R}{V}\right)\left(\frac{R}{P}\right)$$

$$C_P - C_V = \frac{RT^2}{PV} = \frac{R^2T}{RT} = R$$

Hence $\qquad C_P - C_V = R$

[R is in heat units]

For a gas obeying Van der Waals equation

$$\left(P + \frac{a}{V^2}\right)(V - b) = RT$$

$$\left(P + \frac{a}{V^2}\right) = \frac{RT}{(V - b)}$$

$\therefore \qquad \left(\frac{\partial P}{\partial T}\right)_V = \frac{R}{V-b}$

and $\qquad \left(\frac{-2a}{V^3} + \frac{RT}{(V-b)^2}\right)\left(\frac{\partial V}{\partial T}\right)_P = \frac{R}{V-b}$

But $\qquad C_P - C_V = T\left(\frac{\partial P}{\partial T}\right)_V \left(\frac{\partial V}{\partial T}\right)_P$

$$C_P - C_V = \frac{T\left(\frac{R}{V-b}\right)\left(\frac{R}{V-b}\right)}{\left[\frac{RT}{(V-b)^2} - \frac{2a}{V^3}\right]}$$

$$C_P - C_V = \frac{RT\,(V-b)^2}{RT\,(V-b)^2}\left[\frac{R}{1 - \frac{2a(V-b)}{V^3 RT}}\right]$$

Neglecting 6 as compared to V,

$$C_P - C_V = \frac{R}{\left(1 - \frac{2a}{VRT}\right)}$$

Since a is also small as compared to V

$$C_P - C_V = R\left(1 + \frac{2a}{VRT}\right) \qquad \text{...(iii)}$$

[R is in heat units.]

Example 40:

Derive the specific heat relation or show that

$$C_P - C_V = -T\left(\frac{\partial V}{\partial T}\right)_P^2 \left(\frac{\partial P}{\partial V}\right)_T$$

and $\qquad C_P - C_V = -TEa^2V$

where T is the absolute temperature, Sthe bulk modulus of elasticity, a the coefficient of volume expansion and V the specific volume,

Solution:

We have $$C_P\left(\frac{\partial H}{\partial T}\right) = T\left(\frac{\partial S}{\partial T}\right)_P$$

and $$C_V\left(\frac{\partial H}{\partial T}\right)_V = T\left(\frac{\partial S}{\partial T}\right)_V$$

Considering S as a function of temperature and volume,

$$dS\left(\frac{\partial S}{\partial T}\right)_V = dT + \left(\frac{\partial S}{\partial V}\right)_T dV$$

$$\left(\frac{\partial S}{\partial T}\right)_P = \left(\frac{\partial S}{\partial T}\right)_V\left(\frac{\partial T}{\partial T}\right) + \left(\frac{\partial S}{\partial V}\right)_T\left(\frac{\partial V}{\partial T}\right)_P$$

But from equation (v) of article 6.52,

$$\left(\frac{\partial S}{\partial V}\right)_T = \left(\frac{\partial P}{\partial T}\right)_V$$

$$\therefore \quad T\left(\frac{\partial S}{\partial V}\right)_P = T\left(\frac{\partial S}{\partial T}\right)_V + T\left(\frac{\partial P}{\partial T}\right)_V\left(\frac{\partial V}{\partial T}\right)_P$$

$$C_P = C_V + T\left(\frac{\partial P}{\partial T}\right)_V\left(\frac{\partial V}{\partial T}\right)_P \qquad \text{...(i)}$$

Taking the general equation of state for a gas as

$$P = f(V, T)$$

$$C_P = C_V + T\left(\frac{\partial P}{\partial T}\right)_V\left(\frac{\partial V}{\partial T}\right)_P$$

$$\frac{dP}{dT} = \left(\frac{\partial P}{\partial T}\right)_V + \left(\frac{\partial P}{\partial V}\right)_T\left(\frac{\partial V}{\partial T}\right)_P$$

At constant pressure,

$$dP = 0$$

$$\therefore \quad \left(\frac{\partial P}{\partial T}\right)_V = -\left(\frac{\partial P}{\partial V}\right)_T\left(\frac{\partial V}{\partial T}\right)_P$$

Substituting this value in eaquation (i)

$$C_P - C_V = -\left(\frac{\partial P}{\partial V}\right)_T \left(\frac{\partial V}{\partial T}\right)_P^2 \qquad \text{...(ii)}$$

$$C_P - C_V = -T\left(\frac{\partial V}{\partial T}\right)_P^2 \left(\frac{\partial P}{\partial V}\right)_T \qquad \text{...(iii)}$$

But $\quad E = V\left(\frac{\partial P}{\partial V}\right)_T$

and $\quad \alpha = \frac{1}{V}\left[\frac{\partial V}{\partial T}\right]_P$

Substituting thee values in equation (ii)

$$C_P - C_V = -TEa^2V. \qquad \text{...(iv)}$$

Example 41:

Show that $\quad C_2 - C_1 = \frac{dL}{dT} - \frac{L}{T}$

where C_l and C_2 represent the specific heta of a liquid and it saturted vapour and l is the latent heat of the vapour.

Solution:

For a change of state from liquid to vapour,

$$S_2 - S_1 = \frac{L}{T} \qquad \text{...(i)}$$

Here S_1 and S_2 are the entropies in the liquid and vapour states respectively.

Differentiating equation (i) with respect to T

$$\frac{dS_2}{dT} - \frac{dS_1}{dT} = -L.\frac{1}{T^2} + \frac{1}{T}.\frac{dL}{dT}$$

$$T\left(\frac{dS_2}{dT}\right) - T\left(\frac{dS_1}{dT}\right) = -\frac{L}{T} + \frac{dL}{dT}$$

$$C_2 - C_1 = \frac{dL}{dT} - \frac{L}{T}$$

This equation is known an Clausius latent heat equation.

Example 42:

Calculate under what pressure water would boil at 160°C if the change in specific volume when 1 gram of water is converted into steam is 1676 cc. Given latent heat of vaporisation of steam = 540 cal per gram; J = 4.2 × 10⁷ ergs/cal and one atmosphere pressure = 10⁶ dynes/cm².

Solution:

Here $\delta H = 540 \text{ cal} = 540 \times 4.2 \times 107 \text{ erga}$

$\delta V = 1676 \text{ cc}$

$T = 373 \text{ K, and } 150°C = 423 \text{ K}$

$\delta T = 423 - 373 = 50 \text{ K}$

$\delta P = ?$

Applying these values in the Maxwell's thermodynamical relation

$$\left(\frac{\delta H}{\partial V}\right)_T = T\left(\frac{\partial P}{\partial T}\right)_V$$

$$\frac{540 \times 4.2 \times 10^7}{1676} = 373\left(\frac{\partial P}{50}\right)$$

$$\partial P = \frac{540 \times 4.2 \times 10^7 \times 50}{373 \times 1676}$$

$= 1.814 \times 10^6 \text{ dynes/cm}^2$

$= 1.814 \text{ atmospheres}$

Therefore, the pressure at which water would boil at 150°C

$= 1.814 + 1.000$

= 2.814 atmospheres.

Example 43:

Calculate under what pressure ice would freeze at –1°C, if the change in specific volume when 1 gram of water freezes into ice is 0.091 cc.

Given latent heat of fusion of ice = 80 cal/g, J = 4.2 × 10⁷ ergs/cal and one atmosphere pressure = 10⁶ dynes/cm³.

Solution:

Here, $\delta H = 80 \text{ cal} + 80 \times 4.2 \times 10^7 \text{ ergs}$

$\partial V = 0.091 \text{ cc}$

$T = 273 \text{ K}$

$\partial T = 273 - 272 = 1\text{K}$

$\partial P = ?$

Applying these values in the Maxwell's thermodynamical relation,

$$\left(\frac{\delta H}{\partial V}\right)_T = T\left(\frac{\partial P}{\partial T}\right)_V$$

$$\frac{80 \times 4.2 \times 10^7}{0.091} = \frac{273 \times \partial P}{1}$$

$$\partial P = 135.2 \times 10^6 \text{ dyne/cm}^2$$

$$= 135.2 \text{ atmospheres}$$

Therefore, the pressure under which ice would freeze = atmospheric pressure + ∂P = 1 + 135.2 = **136.2 atmospheres.**

Example 44:

Calculate the specific heat of saturated steam given that the specific heat of water at 100°C = 1.01 and latent heat of vaporisation decreases with rise in temperature at the rate of 0.64 cal/K. Latent heat of vaporization of steam = 540 cal.

Solution:

We have $C_1 = 1.01$

$C_2 = ?$

$T = 373 \text{ K}$

$$\frac{dL}{dT} = -0.64 \text{ cal/K}$$

$$C_2 - C_1 = \frac{dL}{dT} - \frac{L}{T}$$

$$C_2 = C_1 + \frac{dL}{dT} - \frac{L}{T}$$

$$= 1.01 + (-0.64) - 540/373$$

$$= \mathbf{-1.077\ cal/g.}$$

It shows that the specific heat of saturated steam is negative.

Example 45:

Calculate the specific heat of saturated steam at 100°C from the following data:

L at 90°C + 545.25 cal

L at 100°C = 539.30 cal

L at 110°C = 533.17 cal

Specific heat of water at 100°C = 1.013 cal/g.

Solution:

Here $\frac{dL}{dT} = \frac{533.17 - 545.25}{20}$

$$= \frac{-12.08}{20} = -0604 \text{ cal/K}$$

$$L = 539.3 \text{ cal}$$

$$T = 373 \text{ K}$$

$$C = 1.013$$

$$C = ?$$

$$C_2 - C_1 = \frac{dL}{dT} - \frac{L}{T}$$

$$C_2 = C_1 + \frac{dL}{dT} - \frac{L}{T}$$

$$C_2 = 1.013 - 0.604 - \frac{539.3}{373}$$

$$\mathbf{C_2 = -1.036\ cal/g.}$$

Example 46:

Using Maxwell's thermodynamical relations, prove that for any substance, the ratio of the adiabatic and isothermal elasticities it equal to the ratio of the two specific heats.

Solution:

From definition,

Isothermal elasticity $E_T = -V\left(\frac{\partial P}{\partial V}\right)_T$

Adiabatic elasticity $E_T = -V\left(\frac{\partial P}{\partial V}\right)_S$

$$\therefore \quad \frac{E_S}{E_T} = \frac{-V\left(\frac{\partial P}{\partial V}\right)_S}{-V\left(\frac{\partial P}{\partial V}\right)_T}$$

$$= \frac{\left(\frac{\partial P}{\partial V}\right)_S}{\left(\frac{\partial P}{\partial V}\right)_T}$$

$$= \frac{\left(\frac{\partial P}{\partial T}\right)_S \left(\frac{\partial T}{\partial V}\right)_S}{\left(\frac{\partial P}{\partial S}\right)_T \left(\frac{\partial S}{\partial V}\right)_T}$$

$$\frac{E_S}{E_T} = \frac{\left(\frac{\partial S}{\partial P}\right)_T \left(\frac{\partial T}{\partial V}\right)_S}{\left(\frac{\partial T}{\partial P}\right)_S \left(\frac{\partial S}{\partial V}\right)_T} \quad \text{...(i)}$$

But from the first four Maxwell s equations

$$\left(\frac{\partial S}{\partial P}\right)_T = -\left(\frac{\partial V}{\partial T}\right)_P$$

$$\left(\frac{\partial T}{\partial V}\right)_S = -\left(\frac{\partial P}{\partial S}\right)_V$$

$$\left(\frac{\partial T}{\partial P}\right)_S = \left(\frac{\partial V}{\partial S}\right)_P$$

$$\left(\frac{\partial S}{\partial V}\right)_T = \left(\frac{\partial P}{\partial T}\right)_V$$

∴ Substituting these values in equation (i)

$$\frac{E_S}{E_T} = \frac{\left(\frac{\partial V}{\partial T}\right)_P \left(\frac{\partial P}{\partial S}\right)_V}{\left(\frac{\partial V}{\partial S}\right)_P \left(\frac{\partial P}{\partial T}\right)_V}$$

$$\frac{E_S}{E_T} = \frac{\left(\frac{\partial S}{\partial V}\right)_P \left(\frac{\partial V}{\partial T}\right)_P}{\left(\frac{\partial S}{\partial P}\right)_V \left(\frac{\partial P}{\partial T}\right)_V}$$

$$= \frac{\left(\frac{\partial S}{\partial T}\right)_P}{\left(\frac{\partial S}{\partial T}\right)_V}$$

$$= \frac{\left(\frac{\partial H}{\partial T}\right)_P}{\left(\frac{\partial H}{\partial T}\right)_V}$$

But $\left(\frac{\partial H}{\partial T}\right)_P = C_P$

and $\left(\frac{\partial H}{\partial T}\right)_V = C_V$

$$\therefore \quad \frac{E_S}{E_T} = \frac{C_P}{C_V} = \gamma .$$

Example 47:

Using MavuxIFs thermodynamical relatwns, prove that the ratio of the adiabatic to the isobaric coefficient of expansion is

$$\frac{1}{(1-\gamma)}$$

Solution:

Adiabatic volume coefficient of expansion,

$$\alpha_2 = \frac{1}{V}\left(\frac{\partial V}{\partial T}\right)_S \qquad \text{...(i)}$$

Isobaric volume coefficient of expansion,

$$\alpha_P = \frac{1}{V}\left(\frac{\partial V}{\partial T}\right)_F \qquad ...(ii)$$

$$\therefore \qquad \frac{\alpha_S}{\alpha_P} = \frac{\frac{1}{V}\left(\frac{\partial V}{\partial T}\right)_S}{\frac{1}{V}\left(\frac{\partial V}{\partial T}\right)_P} \qquad ...(iii)$$

$$= \frac{\left(\frac{\partial V}{\partial T}\right)_S}{\left(\frac{\partial V}{\partial T}\right)_P} = \frac{1}{\left(\frac{\partial T}{\partial V}\right)_S \left(\frac{\partial V}{\partial T}\right)_P}$$

But $\left(\frac{\partial T}{\partial V}\right)_S = -\left(\frac{\partial P}{\partial S}\right)_V$

$$\therefore \qquad \frac{\alpha_S}{\alpha_P} = -\frac{1}{\left(\frac{\partial P}{\partial S}\right)_V \left(\frac{\partial V}{\partial T}\right)_P}$$

$$= \frac{1}{-\left[\left(\frac{\partial P}{\partial T}\right)_V \left(\frac{\partial T}{\partial S}\right)_V \left(\frac{\partial V}{\partial T}\right)_P\right]}$$

$$= \frac{\left(\frac{\partial S}{\partial T}\right)_V}{-\left(\frac{\partial P}{\partial T}\right)_V \left(\frac{\partial V}{\partial T}\right)_P}$$

$$= \frac{T\left(\frac{\partial S}{\partial T}\right)_V}{-T\left(\frac{\partial P}{\partial T}\right)_V \left(\frac{\partial V}{\partial T}\right)_P}$$

But $\quad C_P - C_V = T\left(\frac{\partial P}{\partial T}\right)_V \left(\frac{\partial V}{\partial T}\right)_P$

and $\quad C_V = T\left(\frac{\partial S}{\partial T}\right)_V$

$$\therefore \quad \frac{\alpha_S}{\alpha_P} = \frac{C_V}{-(C_P - C_V)}$$

or
$$\frac{\alpha_S}{\alpha_P} = \frac{1}{1 - \left(\dfrac{C_P}{C_V}\right)}$$

or
$$\frac{\alpha_S}{\alpha_P} = \frac{1}{1 - \gamma} \qquad \text{...(iv)}$$

Example 48:

Using Maxwell's themwdynamical relations, prove that the ratio of the adiabatic to the isochoric pressure coefficient of expansion is equal to

$$\frac{\gamma}{(\gamma - 1)}$$

Solution:

Adiabatic prenxire coefficient of expansion,

$$\beta_S = \frac{1}{P}\left(\frac{\partial P}{\partial T}\right)_S \qquad \text{...(i)}$$

Isochoric pressure coefficient of expanison.

$$\beta_V = \frac{1}{P}\left(\frac{\partial P}{\partial T}\right)_V \qquad \text{...(ii)}$$

$$\therefore \quad \frac{\beta_S}{\beta_V} = \frac{\dfrac{1}{P}\left(\dfrac{\partial P}{\partial T}\right)_S}{\dfrac{1}{P}\left(\dfrac{\partial P}{\partial T}\right)_V} \qquad \text{...(iii)}$$

$$= \frac{\left(\dfrac{\partial P}{\partial T}\right)_S}{\left(\dfrac{\partial P}{\partial T}\right)_V}$$

$$= \frac{1}{\left(\dfrac{\partial T}{\partial P}\right)_S \left(\dfrac{\partial P}{\partial T}\right)_V}$$

But
$$\left(\frac{\partial T}{\partial P}\right)_S = \left(\frac{\partial V}{\partial S}\right)_P \qquad \text{...(iv)}$$

$$\frac{\beta_S}{\beta_V} = \frac{1}{\left(\frac{\partial V}{\partial S}\right)_P \left(\frac{\partial P}{\partial T}\right)_V} \qquad ...(v)$$

$$= \frac{1}{\left(\frac{\partial V}{\partial T}\right)_P \left(\frac{\partial T}{\partial S}\right)_P \left(\frac{\partial P}{\partial T}\right)_V}$$

$$= \frac{\left(\frac{\partial S}{\partial T}\right)_P}{\left(\frac{\partial V}{\partial T}\right)_P \left(\frac{\partial P}{\partial T}\right)_V}$$

$$= \frac{T\left(\frac{\partial S}{\partial T}\right)_P}{T\left(\frac{\partial V}{\partial T}\right)_P \left(\frac{\partial P}{\partial T}\right)_V}$$

$$= \frac{C_P}{C_P - C_V}$$

or
$$\frac{\beta_S}{\beta_V} = \frac{\left(\frac{C_P}{C_V}\right)}{\left(\frac{C_P}{C_V} - 1\right)} = \frac{\gamma}{\gamma - 1} \qquad ...(vi)$$

Example 49:

Using Maxwell's thermodynamical relations, show that

$$\left(\frac{\partial C_V}{\partial V}\right) = T\left(\frac{\partial^2 S}{\partial V.\partial T}\right) = T\left(\frac{\partial^2 P}{\partial T^2}\right)_V$$

Solution:

From Maxwell's relations,

$$\left(\frac{\partial S}{\partial V}\right)_T = \left(\frac{\partial P}{\partial T}\right)_V \qquad ...(i)$$

Differentiating equation (i) with respect to temperature,

$$\left(\frac{\partial^2 S}{\partial T.\partial V}\right) = \left(\frac{\partial^2 P}{\partial T^2}\right)_V \qquad ...(ii)$$

But $$C_V = T\left(\frac{\partial S}{\partial T}\right)_V \qquad ...(iii)$$

Differentiating equation (iii) with respect to volume,

$$\left(\frac{\partial C_V}{\partial V}\right) = T\left(\frac{\partial^2 S}{\partial T.\partial V}\right) \qquad ...(iv)$$

From equations (ii) and (iv)

$$\left(\frac{\partial C_V}{\partial V}\right) = T\left(\frac{\partial^2 S}{\partial T.\partial V}\right) = T\left(\frac{\partial^2 P}{\partial T^2}\right)_V \qquad ...(v)$$

Example 50:

Using Maxwell's thermodynamical relations, show that

$$\left(\frac{\partial C_P}{\partial P}\right) = T\left(\frac{\partial^2 S}{\partial P.\partial T}\right) = -T\left(\frac{\partial^2 V}{\partial T^2}\right)_P$$

Solution:

From Maxwell's relations,

$$\left(\frac{\partial S}{\partial P}\right)_T = -\left(\frac{\partial V}{\partial T}\right)_P \qquad ...(i)$$

Differentiating equation (i) with respect to temperature,

$$\left(\frac{\partial^2 S}{\partial T.\partial P}\right) = -\left(\frac{\partial^2 V}{\partial T^2}\right)_P \qquad ...(ii)$$

But $$C_P = T\left(\frac{\partial S}{\partial T}\right)_P \qquad ...(iii)$$

Differentiating equation (iii) with respect to pressure,

$$\left(\frac{\partial C_P}{\partial P}\right) = T\left(\frac{\partial^2 S}{\partial T.\partial P}\right) \qquad ...(iv)$$

From equations (ii) and (iv)

$$\left(\frac{\partial C_P}{\partial P}\right) = T\left(\frac{\partial^2 S}{\partial T.\partial P}\right) = -T\left(\frac{\partial^2 V}{\partial T^2}\right)_P \qquad ...(v)$$

Example 51:

Show that for a perfect gas

$$\left(\frac{\partial U}{\partial V}\right)_T = 0$$

Solution:

From the Maxwell's first equation,

$$\left(\frac{\partial S}{\partial V}\right)_T = \left(\frac{\partial P}{\partial T}\right)_V$$

$$\frac{1}{T}\left(\frac{\partial H}{\partial V}\right)_T = \left(\frac{\partial P}{\partial T}\right)_V$$

or $$\left(\frac{\partial H}{\partial V}\right)_T = T\left(\frac{\partial P}{\partial T}\right)_V$$

But $$\delta H = dU + PdV$$

$$\therefore \quad \left(\frac{dU + PdV}{dV}\right) = T\left(\frac{\partial P}{\partial T}\right)_V$$

$$\frac{dU}{dV} + P = T\left(\frac{\partial P}{\partial T}\right)_V$$

$$\left(\frac{dU}{dV}\right)_T = T\left(\frac{\partial P}{\partial T}\right)_V - P$$

But from perfect gas equation,

$$PV = RT$$

$$\therefore \quad \left(\frac{\partial P}{\partial T}\right)_V = \frac{R}{V}$$

$$\left(\frac{\partial U}{\partial V}\right)_T = T\left(\frac{R}{V}\right) - P$$

or $$\left(\frac{\partial U}{\partial V}\right)_T = P - P = 0$$

$$\therefore \quad \left(\frac{\partial U}{\partial V}\right)_T = 0.$$

EXERCISES

1. Describe with diagrams the working of an Otto engine and deduce an expression for its efficiency.
2. Prove the thermodynamical relation

$$\left(\frac{\partial H}{\partial V}\right)_T = -T\left(\frac{\partial P}{\partial T}\right)_V$$

and hence show that

$$L = T(V_2 - V_1)\left(\frac{dP}{dT}\right)$$

3. State the two laws of thermodynamics and explain their significance. Derive an expression for the efficiency of a Carnot's engine.
4. Define Kelvin's 'absolute or thermodynamic scale of temperature. Show how this scale agrees with that of a perfect gas scale. Is a negative temperature possible on this scale?
5. Derive the relation between volume and pressure of a gas undergoing adiabatic changes.
6. Derive an expression for the efficiency of an ideal heat engine, working between the temperatures T_1 and T_2K. Prove that the efficiency of a reversible engine is maximum for the given source and sink.
7. Discuss the Second Law of Thermodynamics and the principle of increase of entropy.
8. Explain fully what you understand by entropy. Show that the change in entropy of a substance in a cyclic process is zero.
9. Deduce Clapeyron's equation. How does it explain the effect of pressure on:

   (i) the melting point of solids, and

   (ii) the boiling point of liquids.
10. What is a reversible thermodynamic process? Describe a reversible heat engine and find an expression for its efficiency.
11. Deduce the adiabatic equation for an ideal gas and give Clement and Desormes' method for determining γ.

12. State the Second Law of Thermodynamics. Describe Carnot's cycle and deduce the efficiency of an ideal heat engine.
13. Explain what you understand by thermodynamic scale of temperature. Show that it agrees with an ideal gas scale. What is zero on this scale?
14. Derive Clapeyron's latent heat equation.

$$\frac{dP}{dT} = \frac{L}{T(V_2 - V_1)}$$

15. State and prove Carnot's theorem.
16. Define Entropy. What is its physical significance? Show that the entropy of a perfect gas remains constant in a reversible process but increases in an irreversible process.
17. Prove that

$$\frac{dL}{dT} - \frac{L}{T} = C_2 - C_1$$

What is the physical significance of negative specific heat?

18. Prove that PV^γ = constant in an adiabatic transformation.
19. What do you understand by the absolute scale of temperature? Show that the zero and the size of the degree on this scale and the gas scale are coincident. Is a negative temperature possible on this scale?
20. Prove the thermodynamic relation

$$\left(\frac{\partial H}{\partial P}\right)_T = -T\left(\frac{\partial V}{\partial T}\right)_P.$$

Hence, show that the increase of pressure heats a body that expands with rise of temperature and cools the one that contracts on heating.

21. State Carnot's theorem. Show how Kelvin used this theorem to define a new scale of temperature which is independent of the nature of the working substance. Compare this temperature scale with the perfect gas scale.
22. Describe the absolute scale of temperature. Why is it adopted as the standard scale? What is the meaning of absolute zero on the Kale?

23. Explain the concept of entropy. Deduce the expression for the entropy of a monoatomic gas.
24. Derive an expression for the efficiency of a Carnot's engine in terms of the temperatures of the source and the sink. Show how an absolute scale of temperature can be defined with the help of the ideal Carnot's engine.
25. Deduce Clausius-Clapeyron's equation.
26. Deduce the Clapeyron's latent heat equation from Maxwell's thermodynamical relations.
27. Deduce Maxwell's thermodynamical relations.
28. Show that

(i) $\left(\frac{\partial S}{\partial V}\right)_T = \left(\frac{\partial P}{\partial T}\right)_V$

(ii) $\left(\frac{\partial H}{\partial V}\right)_T = T\left(\frac{\partial P}{\partial T}\right)_V$

(iii) $\left(\frac{\partial H}{\partial P}\right)_T = -T\left(\frac{\partial V}{\partial T}\right)_P$

(iv) $\left(\frac{\partial T}{\partial V}\right)_S = -T\left(\frac{\partial P}{\partial S}\right)_V$

(v) $\left(\frac{\partial T}{\partial P}\right)_S = T\left(\frac{\partial V}{\partial H}\right)_P$

(vi) $\left(\frac{\partial T}{\partial P}\right)_V\left(\frac{\partial S}{\partial V}\right)_P - \left(\frac{\partial T}{\partial V}\right)_V\left(\frac{\partial S}{\partial P}\right)_V = 1$

and (vii) $\left(\frac{\partial P}{\partial T}\right)_S\left(\frac{\partial V}{\partial S}\right)_T - \left(\frac{\partial P}{\partial S}\right)_V\left(\frac{\partial V}{\partial T}\right)_S = 1$

29. Show that for a homogeneous fluid

$$C_P - C_V = T\left(\frac{\partial P}{\partial T}\right)_V\left(\frac{\partial V}{\partial T}\right)_V$$

30. Show from the consideration of Maxwell's thermodynamical relations that

(i) for a perfect gas

$$C_P - C_V = R, \text{ and}$$

(ii) for a gas obeying Van der Waals equation

$$C_P - C_V = R\left(1 + \frac{2a}{RTV}\right)$$

31. Derive the specific heat relation

$$C_P - C_V = -TE\alpha^2 V.$$

32. What do you understand by the internal energy of a system. Slate the first law of thermodynamics and use it to derive a relation between the volume and temperature of a perfect gas undergoing an adiabatic change of state.

33. Deduce Clausius-Clapeyron's equation and discuss the effect of change of pressure on the melting and boiling points of a substance.

34. Deduce the Clausius-Clapeyron's equation for the change of equilibrium pressure of two phases with temperature.

35. Describe Partington's method for the determination of the ratio of the two specific heats of a gas.

36. What are isothermal and adiabatic changes? Derive the equation PV^γ = constant, for an ideal gas.

37. What is meant by a reversible process? Describe Carnot's cycle and show that all reversible engines working between the same two temperatures have the same efficiency.

38. Calculate the work done in a Carnot's cycle of operations. Deduce the efficiency of a Carnot's engine in terms of the temperatures between which it works.

39. Define entropy. State and prove the principle of increase of entropy.

40. Define work scale of temperature. Show that the ideal gas scale and the work scale are identical.

41. Show that

$\left(\frac{\partial P}{\partial T}\right)_V \left(\frac{\partial V}{\partial T}\right)_V$ is independent of the path for a reversible change.

42. Calculate the change in entropy when m grams of a liquid of specific heat S is heated from T_1 K to T_2 K and then converted into vapour without raising its temperature. Latent heat of vaporisation at T_2 is L.

43. Show that

$$\frac{\left[\frac{\partial P}{\partial T}\right]_S}{\left[\frac{\partial P}{\partial T}\right]_V} = \frac{\gamma}{\gamma - 1}$$

44. What is meant by thermodynamical equilibrium? Discuss the equilibrium between a liquid and its vapour and hence deduce Clausius-Clapeyron's equation.

45. Explain the concepts of reversible and irreversible processes. Obtain an expression for the efficiency of a reversible. Carnot engine with a perfect gas as the working substance. Show that for the same pair of working temperatures, the efficiency of all reversible engines is the same and is the maximum value attainable.

46. Show how the second law of thermodynamics enables us to define a scale of temperature independent of the properties of any working substance. How is the scale realised in practice?

47. Derive Maxwell's thermodynamical equations and show that

$$C_P - C_V = -T\left(\frac{\partial V}{\partial T}\right)_P^2 \left(\frac{\partial P}{\partial V}\right)_T$$

48. Describe in detail Clement and Desorme's method of finding the ratio of the two specific heats of air giving the simple theory of the method. What are the objections to the method and what modifications and improvements have been proposed?

49. Establish the relation connecting the temperature and the volume of a gas undergoing adiabatic compression.

50. Establish the latent heat equation,

$$\frac{dL}{dT} - \frac{L}{T} = C_2 - C_1$$

where C_1 and C_2 are the specific heat of the liquid and the saturated vapour respectively.

51. What do you understand by the Kelvin's thermodynamic scale of temperature ? Prove that the ratio of two temperatures on the thermodynamic scale is identical with the ratio of the same two temperatures measured on the perfect gas scale.

52. Explain what you mean by entropy of a substance. Show that the entropy remains constant in a reversible process but increases in an irreversible process.

53. Prove that $\frac{E_S}{E_T} = \gamma$.

54. State the first law of thermodynamics. Express it mathematically and explain its physical significance.

55. Deduce the first latent heat equation. Discuss how the boiling point of a liquid and the melting point of a solid are affected by change of pressure.

56. Describe Carnot's cycle and obtain an expression for the efficiency of an ideal heat engine in terms of temperatures.

57. Derive the relation between the volume and temperature of a perfect gas undergoing adiabatic compression.

58. Explain what you mean by entropy of a substance. Show that for any reversible cyclic change of a system/the total change of entropy is zero. Show that this statement is not true for an irreversible change.

59. Discuss the equilibrium between a liquid and its vapour and hence deduce Clausius-Clapeyron's equation. Explain, how the boiling point of a liquid and the melting point of a solid are affected by change of pressure.

60. Distinguish between isothermal and adiabatic changes. Show that for an adiabatic change in a perfect gas

$$PV^{\gamma} = \text{constant}.$$

61. Explain the terms entropy and isentropics. Show that the entropy of a system tends to a maximum. Represent the Carrot's, cycle on a temperature entropy diagram and prove that its area

represents available energy. Establish directly from the above considerations that the efficiency of the engine is

62. Define entropy. What is its physical significance? Show that the entropy remains constant in a reversible process. What happens in case the process is an irreversible one?

63. Enunciate the second law of thermodynamics. Define a temperature scale without making use of the peculiarities of any selected thermometric substance. Show that Kelvin's work scale is such a scale and that the ratio of the two temperatures as measured on the Kelvin scale is identical with the ratio of the same two temperatures on a perfect gas scale.

64. State the first law of thermodynamics and use it to derive a relation between pressure and temperature of a perfect gas undergoing an adiabatic change.

65. Under suitable conditions the Joule-Thomson effect results in cooling and so does an adiabatic expansion of a gas. Distinguish between these two types of cooling.

66. Deduce the following Maxwell's relations

(i) $\left(\frac{\partial S}{\partial V}\right)_T = \left(\frac{\partial P}{\partial T}\right)_V$

(ii) $\left(\frac{\partial S}{\partial P}\right)_T = \left(\frac{\partial V}{\partial T}\right)_P$.

67. With the help of appropriate Maxwell's relations, show that for a substance

$$C_P - C_V = -TVE\alpha^2.$$

68. State Carnot's theorem and deduce it from the second law of thermodynamics.

69. Derive the Clausius Clapeyron's latent heat equation for first order phase changes.

$$\frac{dP}{dT} = \frac{L}{T(V_2 - V_1)}.$$

70. Derive Maxwell's first and second thermodynamical relations and deduce the Clausius-Clapeyron's latent heat equation from Maxwell's second relation.

71. Define Carnot's cycle. Explain the working of a Carnot's heat engine. Calculate its efficiency when a perfect gas is the working substance.
72. What is meant by the order of a phase change? Discuss it giving examples.

    State and discuss Zeroth law of thermodynamics.
73. Define isothermal and adiabatic transformations. Give examples.

    Obtain equation of state for an ideal gas undergoing (i) an isothermal transformation and (ii) an adiabatic transformation.
74. Dry air at N. T. P. (normal temperature and pressure) is compressed adiabatically to one-third of its original volume. Calculate the resulting pressure. Given that the ratio C_p/C_v of air is 1.40.
75. What is the importance of Clausius Clapeyron latent heat equation? Derive this equation and outline its applications.
76. For a quasi-static adiabatic process of an ideal gas, prove

$$\frac{T}{(P)^{(\gamma-1)/\gamma}} = \text{constant.}$$

77. State Carnot's Theorem and show that it is a necessary consequence of Second Law of Thermodynamics. Prove that the efficiency of a Carnot engine using an ideal gas as a working substance is $\eta = \frac{T_1 - T_2}{T_1}$
78. Derive the following relations:

    (a) $C_P - C_V = -T\left(\frac{\partial V}{\partial T}\right)_P^2\left(\frac{\partial P}{\partial V}\right)_T$

    (b) $TdS = C_P dT - T\left(\frac{\partial V}{\partial T}\right)_P dP$

    (c) $\frac{\left[\frac{\partial P}{\partial T}\right]_S}{\left[\frac{\partial P}{\partial T}\right]_V} = \frac{\gamma}{\gamma - 1}$
79. Write short notes on:

    (i) Isothermal Process

(ii) Adiabatic Process

(iii) Isochoric Process

(iv) Carnot's engine

(v) Carnot's theorem

(vi) Second Law of thermodynamics

(vii) Clement and Desorme'Ls method

(viii) Ruchhardt's experiment for γ

(ix) Absolute gas scale

(x) Rankine cycle

(xi) Diesel engine

(xii) Steam engine

(xiii) Otto cycle

(xiv) Entropy is a measure of disorder

(xv) Entropy tends to a maximum

(xvi) Third Law of Thermodynamics

(xvii) Absolute zero temperature

(xviii} Entropy of a perfect gas

(xix) Temperature Entropy diagram

(xx) Thermodynamic system

(xxi) Thermal Equilibrium

{xxii) Concept of Temperature

(xxiii) Concept of Heat

(xxiv) Zeroth Law in Thermodynamics

(xxv) Phase changes of the second order.

80. A motor car tyre has a pressure of 3 atmospheres at the room temperature of 27°C. If the tyre suddenly bursts what is the resulting temperature?

81. Calculate the temperature at which ice will freeze if the pressure is increased by 135.2 atmospheres. The change in specific volume when 1 gram of water freezes into ice is 0.091 cm^3. One atmospheric pressure = 10^6 dynes/cm^3. Latent heat effusion of ice = 80 cal/g. and J = 4.2×10^7 ergs/cal.

82. Calculate the temperature at which water will boil if the pressure is increased by 1.814 atmospheres. Given that the change in specific volume when one gram of water is converted into steam is 1676 cm^3. Latent heat of vaporization of steam = 540 cal/g. J = 4.2 × 107 ergs/cal and one atmosphere pressure = 10^{37} dynes/ cm^3.

83. Calculate the change in entropy when 20 grams of ice at 0°C is converted into water at the same temperature.

84. Calculate the change in entropy when 10 kg. of water at 100°C is converted into steam at the same temperature.

85. Calculate the increase in entropy when 5 g of ice at –10°C is converted into steam at 100°C. Sp. heat of ice = – 0.5; Latent heat of ice = 80 cal/g; Latent heat of steam = 540 cal/g.

86. Two gram molecules of a gas expands isothermally to four times its initial volume. Calculate the change in its entropy in terms of the gas constant.

87. 100 g of water at 0°C is mixed with an equal mass of water at 80°C. Calculate the resultant increase in entropy.

88. Calculate the change in entropy when 100 g of water at 15°C is mixed with 160 g of water at 40°C. Specific heat of water may be assumed as equal to I.

89. Calculate the change in the melting point of wax for a pressure of 50 atmospheres from the following data. Melting point of wax = 64°C. Specific volume of solid at 64°C = 1.161 and that of the liquid at 64°C = 1.166. Latent heat = 97 cal/g.

90. Calculate the change in temperature of the boiling point of water due to a change of pressure of 1 cm of Hg. L = 536 cal/g. Volume of 1g of water at 100°C = 1 cc and volume of 1 g of saturated steam at 100°C == 1600 cc.

91. Calculate the pressure required to make ice freeze at –1°C. Change in specific volume when one gram of water freezes into ice is equal to 0.091 cm^3. J = 4.2 × 10^7 ergs/ad. One atmosphere = 10^6 dynes/cm^2. Latent heat of ice = 80 cal/g.

92. Calculate the change in entropy when one gram of ice is mixed with 2 millilitres of water at 55°C. Assume that the specific heat and the specific gravity of water remain constant (*i.e.*, equal to

1 in both the cases) between 0°C and 55°C. Latent heat of ice = 80 cal/g.

93. A Carnot's engine takes in 1000 kilocalories of heat from a reservoir at 627°C and exhausts it to a sink at 27°C:

    (a) What is its efficiency?

    (b) How much work does it perform? Express it:

    (i) in ergs,

    (ii) kilo-Watt hours,

    (iii) electron volts.

94. A quantity of air (γ = 1.4) at 27°C is compressed suddenly to 1/4 of its original volume. Find the final temperature.

95. A quantity of air at 27°C and atmospheric pressure is suddenly compressed to 1/5 of its original volume. Find:

    (i) the final pressure, and

    (ii) the final temperature.

96. Find the efficiency of the Carnot's engine working between 150°C and 50°C.

97. Find the efficiency of a Carnot's engine working between 227°C and 27°C.

98. A Carnot's engine whose temperature of the source is 400 K takes 500 calories of heat at this temperature and rejects 400 calories of heat to the sink. What is the temperature of the sink? Calculate the efficiency of the engine.

99. A Carnot's engine is operated between two reservoirs at temperatures of 500 K and 400 K. If the engine receives 2000 calories of heat from the source in each cycle, calculate:

    (a) the amount of heat rejected to the sink in each cycle,

    (b) the efficiency of the engine, and

    (c) the work done by the engine in each cycle in:

    (i) joules,

    (ii) kilo-Watt hours.

100. A Carnot's engine working as a refrigerator between 250 K and 300 K receives 1000 calories of heat from the reservoir at the lower temperature:

(i) Calculate the amount of heat rejected to the reservoir at the higher temperature,

(ii) Calculate also the amount of work done in each cycle to operate the refrigerator.

101. Calculate the depression in the melting point of ice produced by 2 atmospheres increase of pressure. Given latent heat of ice = 80 cal/g and the specific volumes of 1 gram of ice and water at 0°C are 1.091 cm^3 and 1.000 cm^3 respectively.

102. Find the increase in the boiling point of water at 100°C when the pressure is increased by 2 atmospheres. Latent heat of vaporisation of steam is 540 cal/g and 1 g of steam occupies 1677cm^3 volume.

103. Calculate the change in the melting point of naphthalene for 2 atmospheres rise in pressure, given that its melting point is 80°C. Latent heat of fusion is 4563 cal/mol and increase in volume on fusion is 18.7 cm^2/mol. 1 cal = 4.2 × 10^7 ergs.

2

STATISTICAL THERMODYNAMICS

2.1 INTRODUCTION

A physical entity generally manifests itself in (i) a microscopic and (ii) a macroscopic state. We have dealt with macroscopic or the bulk state of matter in the preceding chapters. The macroscopic variables were pressure, volume, temperature, entropy, etc. The macroscopic variables were related mutually on thermodynamical considerations. Further, we assumed that the number of atoms/molecules in them (microscopic details of a macrosystem) were exorbitantly large $\approx 10^{25}$ m^{-3}. The microscopic systems, on the other hand, have very few particles.

Now, we know that the actual systems, the macro systems, are formed of the microsystems and so there should be a definite relation between the two. Classically, if we want to have a complete description of a system we need to know actual velocity (or momentum), location and mutual interaction (if any, with other particles) of every particle. Since every panicle is described by three positions and three momentum coordinates, hence to describe a macrosystem we need a very large number of data. Further, any change in the parameter of an individual particle should be reflected in the properties of the system as a whole.

This poses a question: How to deal with such a vast data which even a super-computer is not capable to handle, if at all we are able to collect it. If, however, we remember that:

(i) much of the data and analytical results obtained on micro level are not useful in the study of the macrosystem, and

(ii) the application of classical dynamical principles to a many particle system is theoretically inapplicable, we can easily dispense with the technically impossible and practically useless classical kinetic and thermodynamic methods in favour of new statistical methods.

Statistics is applied only to such systems which contain a large number of identical individuals which may be distinguishable by means of some property. In a statistical method, we study the behaviour of aggregates of particles to calculate the average value without investigating the detailed behaviour of the system and finding there from the value for each individual. Statistically it is sufficient to know the most probable distribution of particles in different accessible states. This branch of physics has seen its development in the hands of Maxwell, Boltzmann, Gibbs, Fermi-Dirac, Bose-Einstein etc.

In the following sections we will see the importance of the concept of probability in the development of the statistical theory. Hence, we shall first consider some elementary theorems on the calculus of probability.

2.2 PROBABILITY

It refers to a class of events of which the particular event is one member. Mathematically the probability is defined as: When a certain aspect is possible in a ways and is not possible in b ways, then the probability of occurring the aspect is a/(a + b). (Newton defined probability as the ratio of a certain number of aspects, which were possibly observable, to the total number of aspects included in that class.) Here (a + b) is the total number of aspects each of which has the same chance of occurrence.

Now, as we have shown, if an event can occur in a ways, the probability of occurring of this event is a/(a + b), and similarly, if it fails to occur in b ways, then the probability of its failing is b/(a + b). The sum of the two

$$\frac{a}{a+b} + \frac{b}{a+b} = 1$$

corresponds to the certainty of some event, since the event must occur or fail to occur. If b = 0, the probability of occurrence becomes certain but if a = 0, the probability of its failure is certain.

We may define the probability for a certain event occurring m times in n unrestricted (free) attempts as,

$$P = \lim_{n \to \infty} \frac{m}{n}.$$

In the foregoing discussion we have assumed that every event is equally likely.

Calculation of Probabilities

We can extend the theory of probability in order to include the occurrence of two or more independent events. The combined probability in such cases is given by the product of separate probability of each event. Let us examine some cases:

(i) *Tossing of a coin:* On tossing a single coin, there are two possible aspects that may be presented by a fallen coin, a "head" or a "tail". We may assume here that there is no preference for any particular event, so that after making a large number of throws, the ratio of the number of heads to the number of tails is unity. Thus, for a normal coin the probability of each event is 1/2.

(ii) *Tossing of two unlike coins*: Suppose we toss two coins of different denominations or of different sizes, or of different colours, we have four complexions.

1. $Head_1$ $Head_2$
2. $Head_1$ $Tail_2$
3. $Tail_1$, $Tail_2$
4. $Tail_1$ $Head_2$.

Thus, the probability of any one complexion on a single toss will be 1/4 = 1/2.1/2, Or the composite probability is the product of probability of individual events.

(iii) *Tossing of two similar coins*: If two coins cannot be distinguished from each other, then complexions 2 and 4 of (ii) are indistinguishable whence the probability of one head and one tail is twice than that of two heads or two tails.

(iv) *Tossing of two coins in general*: Let us represent occurrence of heads by A_1, A_2 and the occurrence of tails by B_1, B_2 when two coins are tossed. Then the composite event can be represented by $(A_1 + B_1)(A_2 + B_2) = A_1A_1 + A_1B_2 + A_2B_1 + B_1B_2$

The RHS can be tabulated as under .

S.N.	Statistical State	Symbol	Statistical Weight	Probability
1.	Both Heads	A_1A_2	1	1/4
2.	Both Tails	B_1, B_2	1	1/4
3.	One Head and one Tail	A_1B_2 or B_1A_2	1 + 1	2/4

(v) *Tossing of three unlike coins:* With three unlike coins, we have three distinguishable heads A_1, A_2, A_3 and three distinguishable tails, B_1, B_2, B_3. Here we can have eight equally probable complexions $A_1 A_2 A_3$. $A_1 A_2 B_3$, $A_1 B_2 B_3$, $B_1 A_2 B_3$, $B_1 A_2 B_3$, $B_1 B_2 A_3$ and $B_1 B_2 B_3$ which are really obtained from the product of $(A_1 + B_1)$, $(A_2 + B_2)$, $(A_3 + B_3)$.

(vi) *Tossing of three similar coins:* In this case we have only four probable states with complexions

1. All heads.
2. Two heads and one tail
3. All tails.
4. One head and two tails.

We can tabulate the result as in (iv) for

$(A + B)(A + B)(A + B) = A_3 + 3A^2B + 3AB^2 + B^3$.

S.N.	Statistical State	Symbol	Statistical Weight	Probability
1.	All heads	A.A.A	1	1/8
2.	Two heads one tail	A.A.B	3	3/8
3.	One head two tails	A.B.B	3	3/8
4.	All tails	B.B.B	1	1/8

(vii) *Tossing of n coins*: Suppose all the coins are different. Then the possible complexions (also known as probability number) will be $(A_1 + B_1)(A_2 + B_2)$ ..... $(A_n + B_n)$. And, if the coins are all alike the total number of possible complexions will be

$$(A + B)^n = A^n + {}^nC_1A^{n-1} B + {}^nC_2 A^{n-2} B^2 + ... {}^nC_rA^{n-r}B^r + ... + {}^nC_n B^n$$

$$\text{or} \quad = \sum_{r=0}^{r=n} C_r A^{n-r}B^r$$

$$\text{or, also } (A + B)^n = \sum_{r=0}^{r=n} C_r A^rB^{n-r}$$

$$\equiv \sum_{r=0}^{r=n} C_r A^{n-r}B^r$$

where $^nC_r = \dfrac{n!}{r!(n-r)!} \equiv$ Number of events with r heads/tails.

Now, since A and B indicate one event each, the total number of events are $(1 + 1)^n = 2^n = {}^nC_1 + {}^nC_2 + ... + {}^nC_n$

∴ The probability of r heads and (n – r) tails to appear is

$$P(r, n-r) = \frac{^nC_r}{2^n} = \frac{n!}{r!(n-r)!} \cdot \frac{1}{2^n}$$

Maximum and minimum probability: To find out which of the combination of heads and tails is most likely in a large number of throws of n coins, we have to find the maximum value of ^nC_r. It is known from algebra that the value of ^nC_r is greatest for r = n/2 when n is even and r = n+1/2 when n is odd.

$$P_{max} = \frac{^nC_{n/2}}{2^n} = \frac{n!}{(n/2)!(n/2)} \cdot \frac{1}{2^n}$$

Similarly, the minimum probability, P_{min} obtained for r = 0 or r = n as $^nC_n = 1$.

whence $P_{min} = 1/2^n$

(viii) *Tossing of a single coin n times*: When we toss the same coin second, third,... nth time, its behaviour is again unpredictable. It may show a head or a tail each time with 1/2 probability. Hence a single coin tossed n times (and of course, making large number of such attempts, if we want to have satisfactory results) is just equivalent to the case of tossing of n coins as discussed in (vii).

Another equivalent illustration can be: Take 2 boxes, identical in all respects, and throw n balls from a great distance. Assuming that each ball will fall into one or the other box, the distribution of balls in the boxes will be governed by the relation (vii),

(ix) *Tossing of a dice*: If we toss of dice, having six faces, the probability for each face showing up will be 1/6. Since now the total number of aspects is 6 and only one can show up at a time with equal probability. Again, we may take 2, 3,.... n dices and throw them and calculate the composite probability for a certain complexion.

(x) *Tossing of n dices each having m faces*: Here let us assume that dice 1 has faces a_1, a_2, a_3... a_m; dice 2 has faces b_1 b_2 b_3 ... b_m; dice 3 has faces c_1, c_2, c_3 ; etc., then the possible complexions with n dices will be the sum of each aspect

a_1, a_2, a_3... a_m, v_1, b_2, b_3 ... b_m , c_1, c_2, c_3 ... n sets and, if the dice are all alike, we have

$m_{.a}$, $m_{.b}$, $m_{.c}$... n sets.

If the probability for occurring each of the face is same, *i.e.*, a = b = c... = one aspect, we get total number of possible aspects

$$= \text{m.m.m.... n times} = m^n$$

One can show that the most probable distribution corresponds to equitable distribution of faces *i.e.*, face 1, face 2, face 3, etc., will appear in equal number. Then all the dices showing face 1 have the probability of only $1/m^n$.

A similar illustration to this is : Take *m identical boxes and throw n* balls onto them from a great distance. We will take up this problem later in more detail. As in actual analysis of physical problems we have a large number of particles which are distributed into a large number of (hypothetical) boxes.

(xi) *Probability with weightage*: So far we have considered the case in which the probability of occurring an event and its failure is the same. But there can be cases where the probability of occurring a particular event is greater than that for any other event.

Let a be the probability of occurring an event and b be the probability of its failure, per trial. Then, if the trial is repeated n times, the probability of r favourable events is

$$a^r.b^{n-r} = a^r (1 - a)^{n-r}$$

But there are ^nC_r ways for r favourable events, therefore, the probability of occurring r events is

$$P(r, n - r) = {}^nC_r\, a^r\, (1 - a)^{n-r}$$

Thus, for example, in a dice a = 1/6

$$\therefore \quad P(r, n - r) = {}^nC_r \left(\frac{1}{6}\right)^r \left(\frac{5}{6}\right)^{n-r}$$

But, if certain number, say 3, has a = 2/7 then in n tosses the number 3 will occur r times with a probability of

$${}^nC_r \left(\frac{2}{7}\right)^r \left(\frac{5}{7}\right)^{n-r} \quad \text{etc.}$$

(xii) *Postulate of equal a priori probability:* It states that all accessible microstates corresponding to a macrostate are equally probable. Thus, the postulate claims that the probability of finding a molecule in one region of phase space is identical with that for any other region of equal volume provided the regions are obeying other conditions of equality (*i.e.*, the regions have extensions of same magnitude and correspond to the same energy). Thus, it follows that the probability of occurrence of a macrostate is proportional to the number of microstates associated with it

i.e., $\qquad P_i \propto \Omega_i$

where P_i is the probability and Ω_i is the total number of accessible microstates, in the ith set.

15.3 SOME USEFUL DEFINITIONS

It would be appropriate to discuss, at some length, the physical entities involved, in understanding the statistical problems.

A Physical Quantity

Essentially, we have two kinds of physical quantities (a) the intensive quantities that do not depend on the mass of the system, *e.g.*, P and T, and (b) the extensive quantities which depend on the mass of system, *e.g.*, S, V, E, etc.

System

It is a finite region in the space where the experimental body is kept. It Is obligatory for the system to have a finite mass. The system can be a closed one (may have a real physical boundary) or may have an imaginary boundary in space. Hence it can be penetrable or Impenetrable to the transport of energy. The properties of the particles in the system define the properties of the system.

Macrostate

It is that state of the system which can be perceived by experiments. It is an isolated state of the system (not influenced by external agencies) and can be defined by certain value of pressure P, temperature T, and volume V. A macrostate can also be defined in terms of its microstructure, *e.g.*, by the specification of number of molecules (or phase-points in each cell of the phase space). Let us consider the distribution of 8 particles in two boxes. We can have (0, 8), (1, 7), (2, 6)... (7, 1) and (8, 0) distributions. These 9 distributions form 9 macrostates. Here we may note that a macrostate does not distinguish individuality of particles. Hence, for n particles we have (n + 1) macrostates when the distribution of particles is restricted in 2 boxes. This number will increase if the number of boxes is increased.

Microstate

Each macrostate can have several microstates associated with it. Microstate is that state of the system where the individuality of the particles is not ignored. Hence the microstates are affected by change in position, momentum, energy (and mutual interaction) of individual particles. These states, however, cannot be identified by any physical experiment. Following table will help in understanding the difference between macro and microstates. We take two boxes A and B and four distinguishable particles a, b, c, d, and investigate their distribution.

It is evident from part B of the table that when the individuality of the particles is ignored, there remains no difference between the number of macrostates and associated microstates. By definition, given above, such a distribution will not give statistical probability or correct probable distribution (last column). Another illustration for the above table is to distribute 4 distinct books on 2 shelves. We have a total of 16 distributions as in part A of the table but if we distribute 4 identical books on same 2 shelves we have only 5 arrangements.

From the table, we see that for the most probable distribution (2, 2) we have 6 microstates. In other words, the macro distribution is most probable when number of microstates involved is maximum. In actual physical problems, we deal with extremely large number of particles as $\approx 10 m^{-3}$ whence the number of microstates can be unimaginably large.

Phase Space

Let us consider a particle with only a motion of translation. Then, classically we need three positional and three momentum coordinates to describe its state, (see point P in Fig. 2.1). Just as the space is represented by a three dimensional positional coordinate system, we can also represent momentum space by a three-dimensional momentum coordinate system. The state of the system then can be defined by six-dimensional positional-space and momentum space (phase) system. The two spaces together, defined by 6-dimensional (3 space and 3 momentum) hypothetical coordinate systems, is called phase space. Phase-space is purely a mathematical concept and cannot be depicted by a diagram like Fig. 2.1. So, phase-space is just a hypothetical space occupied by an infinite set of such six coordinate points.

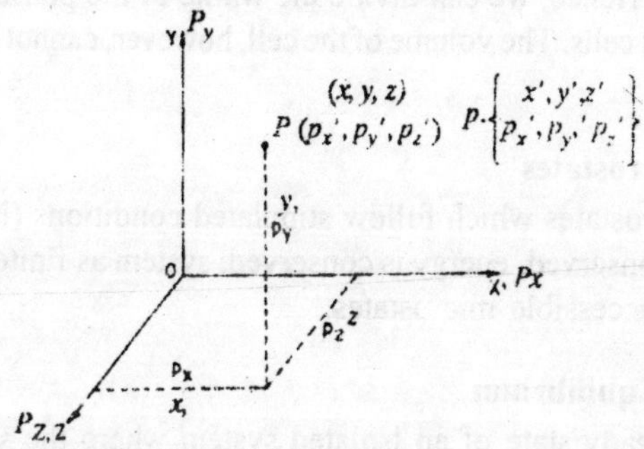

Fig. 2.l.

With the passage of time the point P (known as phase-point) traces out a certain trajectory known as phase-path. As the position and momentum, in translated motion, change with lime, the phase-path may be quite complicated. Such 6-dimensional (6-d) phase space is also known as the μ-space. A large number of 6-d cells, each with a volume of $d\tau = (dx\ dy\ dz)\ (dp_x\ dp_y\ dp_z) = dV.\ dp$, together form the μ-space.

Let us now pass on to a many particle system. Let the system has N particles. Then, to define their state we need 3N positional and 3N momentum coordinates in a 6-d phase-space. We can consider this problem in another way:

A single point in 3-d space represents 3 coordinates; in 6-d space represents 6 coordinates; in n-d space represents n coordinates. Hence in 6N-d space a single point corresponds to 6N coordinates. If N be the number of molecules in a kilogram-mole (or a Mole), the 6N-dimensional phase-space is called r-space (or y-space). We can generalise this representation by assuming that the system has N particles each having f degrees of freedom the Γ- space will then be 2Nf dimensional, having Nf positional coordinates (q_i, i = 1 to Nf) and Nf momentum coordinates (P_i, i = 1 to Nf).

Phase-cells

The concept of phase space enables us to understand the state of micro particles A their position and energy. When a particle changes its momentum and position with time, there will be infinite points in the phase-space which will express the state of this moving particle. Again in a finite volume of phase space, we have infinite microstates available for the particle. Hence, we can divide the whole of the phase space into finite sized small cells. The volume of the cell, however, cannot indefinitely be small.

Accessible Microstates

Those microstates which follow stipulated conditions (like number of particles is conserved, energy is conserved, system as finite extension) are called the accessible mic ɔstates.

The State of Equilibrium

It is the steady state of an isolated system where the values of P, V and T do not change with the passage of time. In such systems, no energy is given to or taken out of it. In statistical terms, equilibrium state corresponds to that macrostate which has maximum microstates. One must, however, remember that an isolated system, in equilibrium, is equally likely to be in any of its accessible microstate.

Events

When an operation is performed or is allowed to take place, the occurrence of certain complexion, configuration or arrangement of the system, at one such trial is called an event. When number of trials is increased the probability of certain event may increase.

2.4 STATISTICS APPLIED TO MOLECULES

Let us consider a monatomic perfect gas. The state of each molecule of the gas in the μ-space is represented by a point (and by a cell in quantum mechanical terms). But, since the gas molecules are in constant motion, their state will be changing continuously in space and, due to collisions (and also mutual interactions), there is a change .in momentum state also. The trajectory of such a molecule (see Fig. 2.2) is given by a constant energy surface given by

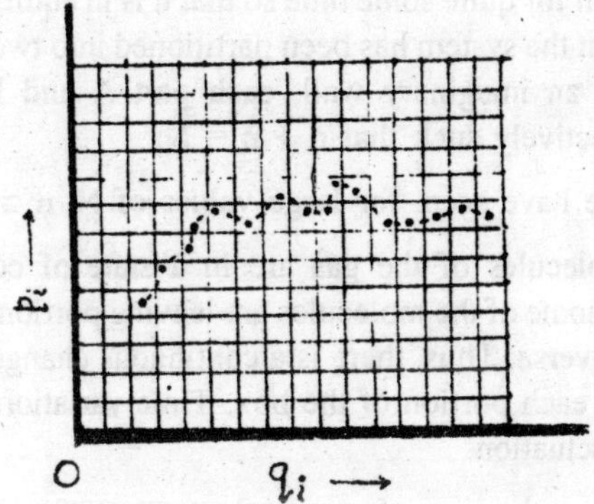

Fig. 2.2

$$E = \frac{p_x^2 + p_y^2 + p_z^2}{2m} = V(q_x, q_y, q_z)$$

The real gas molecules, however, are endowed with mutual interactions in addition to usual collisions, hence their state is better represented by a point in y-space. The rate of change of its state, then, depends on the quantum of interactions. The path of representative point in y-space is given by the relation

$$E = \sum_i T_i + \sum_i V_i + \sum_{i,j}{}' U_{ij}$$

where T_i, V_i and U_{ij} refer to kinetic energy, potential energy and energy of interaction of ith particle. The prime on the summation sign suggests that i = j is not allowed (which would mean self-interaction).

With this background and remembering that the number of molecules is very large in a macrosystem, we now proceed to investigate the distribution details.

2.5 CHARACTERISTIC FEATURES OF MACROSCOPIC SYSTEMS

Let us now discuss some of the characteristic features of the macrosystems. In doing so we will assume that it has a large number of microscopic particles.

Fluctuation

Let the isolated system has N molecules of a perfect gas and that it was left to itself for quite some time so that it is in equilibrium. Further, also consider that the system has been partitioned into two identical (also equal) parts by an imaginary wall, each part A and B has n and n molecules respectively such that n + n = N.

Now, as we have seen, for large values of N, $n \cong n^1 = N/2$.

But the molecules of the gas are in a state of constant random motion. So that some of the molecules are leaving portion .A and moving into B and vice-versa. Thus, there is a continuous change in the number of molecules in each portion of the box. Time variation of this change is known as fluctuation.

Fig. 2.3 shows the state of normal fluctuation of an isolated system. Usually, N/2 – n is only a small number. For large deviation, the number of microstates (and hence the probability) reduce drastically. The number of microstates for different values of N and N/2 – n. For the sake of simplicity the deviation N/2– n is represented as a fraction of total number of particles. Value of R ($\cong P_x/P_{max}$) is also shown to arrive at qualitative results;

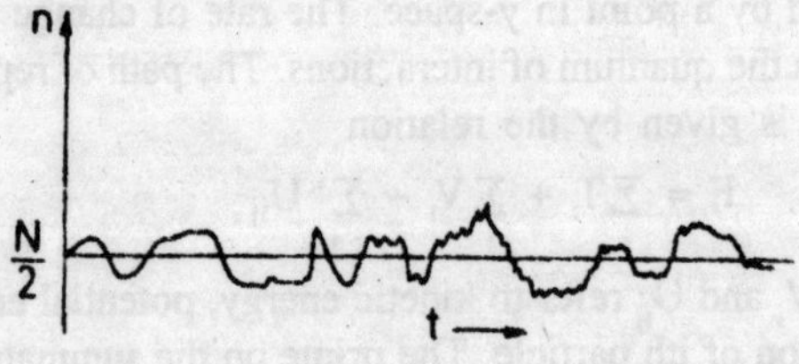

Fig. 2.3

In actual physical problems the number of molecules are $\sim 10^{24} m^{-3}$ whence even 0.01% deviation also becomes quite improbable. This is the reason for quick diffusion and most equitable distribution of air molecules in the available space.

Equilibrium State of the System

The macrostate of a gas at a given instant is obtained from the distribution of molecules at that instant. But, if the macrostate of the multimolecular system does not change with time, the isolated system is said to be in equilibrium. Statistically, in the equilibrium state, the probability of the isolated system being in different accessible states does not vary with time. It may be noted that every system has an inherent tendency to approach equilibrium. If, anyhow, a situation (like point t_1 in Fig. 2.4) is created so that the system goes into a higher fluctuation state (lower probability), the motion of the molecules will set in such a way as to bring the system back into normal fluctuation state, with the passage of time. The time taken by the system to attain normal state from its disturbed (excited) state is called the relaxation *time of decay*. It is very easy to perturb the normal state of the system by external effects (by pressure/wind; diffusion of perfume is one such example). But the system will again tend to attain uniform distribution/most probable distribution. To check whether the macrostate is in equilibrium we plot fluctuation versus time curve for a certain period τ. The experiment is repeated again and again. If the curves agree macrostate will be in equilibrium state. In other words, the time average of number of molecules in certain region of space at equilibrium

$$\overline{n(t)} = \frac{1}{\tau}\int_t^{t+\tau} n(t)dt$$

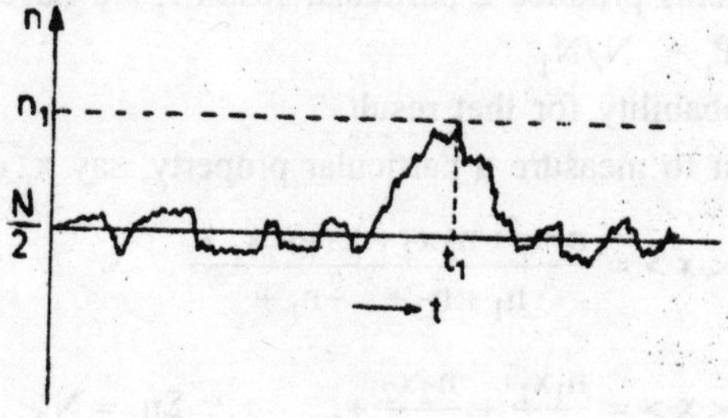

Fig. 2.4

2.6 GIBBS ENSEMBLE

The concept of ensemble was put forth by Gibbs in 1902. In his view, it was a mathematical device which provided us with an opportunity to follow a probabilistic approach in defining the state of the system at a given time. Consider a system A such that a number of similar observations can be taken on it. Further, we assume that the system undergoes exactly similar changes in its state during each observation, *i.e.*, after every set of operations, the system returns back to its initial state (*i.e.*, the system is perfectly reversible). If this be the case, we can define the state of the system in terms of its statistical probability. But this is not practically feasible as:

(i) The number of molecules/particles in the system are so large that it is not possible to have the macrostate exactly the same as it was during the preceding observation,

(ii) Each macrostate has several microstates. We cannot ascertain their identity,

(iii) Each observation may take quite a long time whence for applying statistics, several observations may need several thousand years,

(iv) To manage and process a large number of data, over years together is not possible,

(v) The system may not show reversibility to different operations.

Ensemble was a satisfactory solution to it. An ensemble is a group of N_T identical systems. These N_T systems have the same configuration, energy and accessible states. In other words, the ensemble has N_T isolated systems each of which are exact copies of the experimental system A. Thus, we have a large number of systems that behave exactly in the same manner during a set of operations as the system A. Then, if N_r systems out of N_T systems produce a particular result r, we have

$$P_r = N_r/N_T$$

as the probability for that result.

If we want to measure a particular property, say x, of the system

$$< x > = \frac{n_1x_1 + n_2x_2 + ... + n_rx_r +}{n_1 + n_2 + ... + n_r +}$$

or $$< x > = \frac{n_1x_1}{N_T} + \frac{n_2x_2}{N_T} + ... \qquad \because \Sigma n_i = N_T$$

$$\text{or} \qquad = P_1x_1 + P_2x_2 + ...P_rx_r + ...$$

$$\textit{i.e.,} \qquad = \Sigma P_rx_r$$

where x_r is the property of rth group of N_r similar systems and P_r is the probability for that result.

In this unique way Gibbs replaced the method of time averages (observations made on the same system at different times to obtain average) by ensemble average as shown above. The main advantage of the ensemble method is "the observations on all systems can be taken simultaneously under similar conditions", *e.g.*, test of quality of soaps of different brands will be quicker by millions of users.

2.7 SOME WELL-KNOWN ENSEMBLES

It would be appropriate to classify the ensembles according of their dependency on macroparameters. Since an ensemble is an assembly of macroscopically identical but mutually independent systems, a change in macroparameter (like volume, pressure, energy, number of molecules, etc.,) may lead to another set of systems:

(i) Uniform Ensemble

Those ensembles which have same density in phase space, are called *uniform ensembles*.

Hence, if ΔN represents the number of system, in volume element $d\Gamma$ we have, for a uniform ensemble,

$$\lim_{d\Gamma \to 0} \frac{1}{d\Gamma} \frac{\Delta N}{N} = \text{constant}$$

(ii) Microcanonical Ensemble

In these ensembles the energy of their member systems remain constant, *i.e.*, they do not interact with each other or with the surroundings. Thus, they are free from external effects. The ensemble is thus conceptually an idealised one and can be represented by a function that depends only on the energy. Fig. 2.5 shows such an ensemble, here ε and V are constant. Every member is separated from the other by rigid, impermeable and insulated walls. The density of phase points within each member system is uniform and is zero at the walls. Physically, we may view the microcanonical ensemble as a perfect insulating enclosure encompassing several isolated systems. It is suitable for ideal gases.

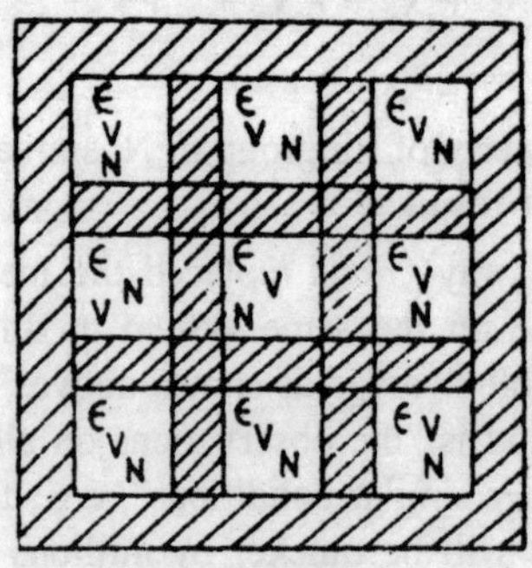

Fig. 2.5

(iii) Canonical Ensemble

In this ensemble all its members are identical and remain in thermal equilibrium with the source (*i.e.*, members of the ensemble can exchange thermal energy with the reservoir). Thus all the members of a canonical ensemble have the same number of particles N, volume V and are at the same temperature T. Fig. 2.6 shows a canonical ensemble. Its members are shown separated by impermeable but conducting rigid, walls and they are in thermal contact (not shown) with the source. Gibbs gave a relation

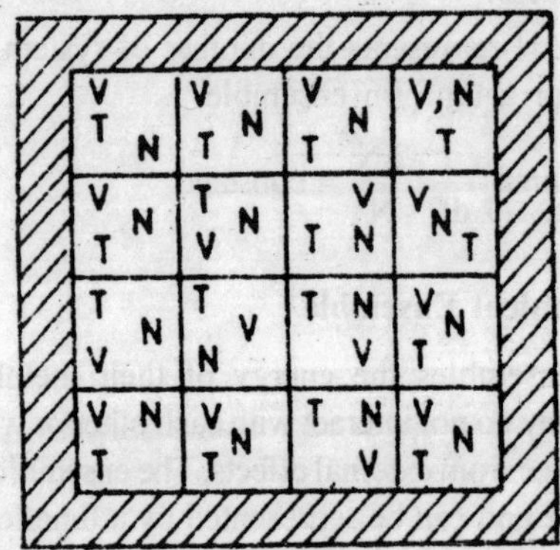

Fig. 2.6

$$\rho(\varepsilon) = Ae^{-\varepsilon\tau}$$

for density of its phase points. A and T in the relation are constants. The ensemble can be viewed physically as a freezer containing water filled metal-ice cube tray.

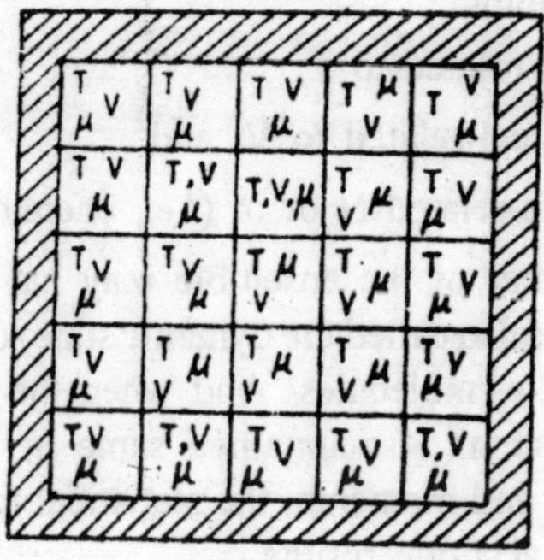

Fig. 2.7

(iv) Grand Canonical Ensemble

The members of this ensemble can interact thermally, mechanically and also diffusively with each other and with the reservoir. Hence, they are at the same temperature of same volume and at same chemical potential. Fig. 2.7 shows such an ensemble. Each member here is separated by flexible, permeable and conducting walls, which help exchange mass and energy between its members and with the surroundings (not shown). The constituent molecules in the member systems are thus of different kinds (different chemical species). The density of phase points in such ensemble is

$$\rho(N) = e^{(\Omega + N\mu - \varepsilon)/\tau}$$

where Ω, μ and τ are constants. A grain of sugar in boiling water can best be viewed as a grand canonical ensemble.

2.8 ENSEMBLE METHOD

The *ensemble average* method is superior to the method of *time average*, in determining average value of a physical quantity. So first we recall the characteristics of an ensemble and its way of representation.

1. It is a group of large number of systems.
2. Every member of an ensemble is an exact miniature copy of the system under test.
3. Every member of an ensemble is subjected to the same experimental conditions.

4. There may be three kinds of interactions between the members of the ensemble:

   (a) Thermal-related to T,

   (b) Mechanical-related to V, and

   (c) Diffusive-related N or m (*i.e.*, chemical potential).

5. Every member of the ensemble may have a large number of microstates (also called the dynamic state) due to random motion of its particles/molecules/ And when the configuration of the member systems is maintained same in terms of generalised coordinates and momenta, the statistical method is expected to yield quite accurate results.

Ensemble Representation

Ensembles can be represented in a Γ-space by the distribution of its phase-points. Generally, the distribution of phase-points in an ensemble is continuous and can be represented by a density function ρ(p, qt). The number of phase-points in the vicinity of (p, q) in Γ-space is given by p (p, q t) dT where d r = dp dq = volume element of Γ- space.

If ρ (p, q t} is known at time t_1, its value can easily be obtained for some other time t_2. Now, from the probability theory

$$d\rho\,(p, q)\ \alpha\ d\Gamma \qquad \therefore d\rho\,(p, q) = \rho(p, q)d\Gamma$$

Here ρ (p, q) is known as the distribution function. The ensemble average of any function R is then given by

$$\langle R\rangle = \int R\,(p, q)\ \rho(p, q)d\Gamma.$$

Selection of Ensemble

(i) Molecules of real gases are usually formed of 2, 3.... atoms whence the density of phase points is not uniform,

(ii) The density of phase points is non-zero for every range of energy,

(iii) Total number of particles and energy is conserved. Point (i) eliminates use of uniform ensemble. Point (ii) eliminates use of microcanonical ensemble. Point (in) eliminates use of grand canonical ensemble to real gases.

But since the properties of these gases such as N, V, T, ε, H etc., are constant the canonical ensemble finds use in description of real gases for their thermodynamic properties.

2.9 STATISTICAL POSTULATES

Following postulates are made for applying the statistical method to physical systems:

(i) Every system has a large number of particles/molecules which are in constant random motion.

(ii) The motion of particles/molccules is described in 2Nf-dimensional phase-space.

(iii) The phase-space is divided into cells.

(iv) Postulate of equal a priori probability holds good.

(v) Probability of a particular configuration is given by the ratio of microstates corresponding to that configuration to total accessible microstates, *i.e.*,

$P_i = \Omega_i/\Sigma\Omega_i$ and the average value of any physical parameter is given by

$$< x > = \Sigma x_i P_i = \frac{\Sigma x_i \Omega_i}{\Sigma \Omega_i}$$

where Qi is the frequency of the configuration.

(vi) The equilibrium state of the system is the state of maximum probability.

(vii) The law of conservation of mass is generally valid whence

$$n_1 + n_2 + n_3 + ... = \Sigma n_i, = N = \text{constant.}$$

(viii) The law of conservation of energy governs the distribution of particles among cells. Thus, if ε_1, ε_2, ε_3 ... denotes energies of each particle in cell 1, 2, 3, etc., then

$$E = \varepsilon_1 n_1, + \varepsilon_2 n_2 + \varepsilon_3 n_3 + ... = \Sigma \varepsilon_i n_i = \text{constant.}$$

(ix) Every point in the phase-space (cell) represents a microstate, the cell volume being h where/is the degree of freedom.

(x) The number of microstates in the energy interval e and $\varepsilon + d\varepsilon$, per unit volume is represented as $\Omega(\varepsilon)$.

(xi) A collection of accessible microstates form a statistical ensemble.

Note: A complexion/configuration/arrangement represents distribution of molecules in different cells, and their form is governed by constraints stated in (vii) and (viii).

2.10 ACCESSIBLE MICROSTATES

Let us now count the number of energy states available in a macrosystem. Let E be the total energy of this system which has been divided into small energy intervals δ E so that a sufficient number of quantum stales are available in each interval of δ E. Now, if Ω (E) is the number density of such states in the energy interval E and E + δE then

$$\Omega(E) = \rho(E)\ \delta E \qquad ...(i)$$

where ρ (E) is density of states (the number of states in unit energy interval). Thus

$$\Omega(E) \propto \delta E \qquad ...(ii)$$

Let the value of Ω(E) changes with the magnitude of energy interval. Hence we may consider Ω(E) as a smoothly varying function of E.

Again, if ϕ(E) denotes the number of states with energy up to E and ϕ(E + δE) denotes the number of states with energy up to E + δE then

$$\Omega(E) = \phi(E + \delta E) - \phi(E) \equiv \frac{\delta\phi}{\delta E}\delta E \qquad ...(iii)$$

Comment on ρ (E). Relations (i) and (ii) (possibly) indicate that ρ(E) is a constant. But it is not so. In fact ρ (E) is an increasing function of E. Let us see how! We know that the total number of microstates is equal to the product of thermodynamic probabilities of different macrostates (composite probability is obtained, always by multiplying probabilities for different complexions), *i.e.*,

$$\Omega = \Omega_1\, \Omega_2\, \Omega_3 \ldots \Omega_i \ldots \Omega_f$$

or $$\Omega = \prod_1 \Omega_i\ (i = 1 \text{ to } f)$$

where f is the degree of freedom.

But Ω_i should be somehow related to E, *i.e.*,

$$\Omega \propto E^a \quad \therefore \Omega \propto E^{fa} \qquad ...(iv)$$

usually a $\cong$ 1/2 and/ranges between 10^{22} and 10^{28}. This suggests that for a certain energy interval δE, ρ(E) increases with the value of E.

2.11 STATISTICAL MECHANICS

Statistical Mechanics deals with systems consisting of many particles and the methods employed are to get a collective or macroscopic property of the system without taking into account the individual motion of the particles The word particle is used in a broad sense meaning a fundamental particle *e.g.*, an electron, an atom, a molecule etc. The particle is a well defined and a stable unit of a given physical system. As it is not practically possible to determine the property of each particle individually, a statistical approach is made by using the concept of probability of distribution. This statistical approach helps in determining the bulk or macroscopic property of the system as a whole. The idea of probability does not imply that the particles move in a random way without obeying the laws. For a many particle system, the statistical analysis is valid regarding the applications of probability distribution of particles.

2.12 STATISTICAL EQUILIBRIUM

Consider that a thermodynamical isolated system consists of N particles. The energy states available to the particles are E_1, E_2, E_3 etc. These energy states may be quantized or may be continuous and are due to vibrational and rotational energy of the particles.

Suppose, that at any given instant of time, n_1 particles are in state of energy E_1, n_2, particles with energy state E_2 and so on.

The total number of particles in the system is,

$$N = n_1 + n_2 + n_3 + ... = \sum_i n_i \quad ...(i)$$

Here $i = 1, 2, 3$ etc.

The total energy of the system,

$$U = n_1E_1 + n_2E_2 + n_3E_3 + ...$$

$$U = \sum_i n_i E_i \quad ...(ii)$$

Equation (ii) refers to the total energy of a system in which the particles are non-interacting. Here, the energy of each particle depends only on the coordinates of the particle in the system. For an isolated system, the total energy U is constant.

For isolated system.

$$U = \sum_i n_i E_i = \text{constant} \quad ...(iii)$$

Consider a gas having N molecules at a certain temperature and pressure. Its volume, temperature and pressure are kept constant *i.e.*, the system is isolated. The total energy of this system remains constant. But, the molecules of the gas collide with each other and also with the walls of the container. Consequently, the number of molecules change from one energy state to the other energy state. It means that the value of n_1, n_2, n_3 etc., continuously change. It can be reasonably assumed that for each microscopic State of a system of particles, there is a particular most favoured distribution. When this distribution or partition is reached, the system attains statistical equilibrium. For an isolated system, the values of n_1, n_2, n_3 etc., vary only near the values corresponding to the most probable distribution. Hence, the basic problem in statistical thermodynamics is to obtain the most probable distribution law for a given composition of the system.

In practice, three most probable distribution laws are used. They are:

(i) Maxwell-Boltzmann Distribution Law.

(ii) Fermi-Dirac Distribution Law (Statistics).

(iii) Bose-Einstein Distribution Law (Statistics).

2.13 PROBABILITY THEOREMS IN STATISTICAL THERMODYNAMICS

The following are the important probability theorems commonly used in statistical thermodynamics:

(1) The number of ways in which N distinguishable particles can he arranged in order is equal to

$$N!$$

(2) The number of different ways in which N particles can be selected from N distinguishable particles irrespective of the order of selection is equal to

$$\frac{N!}{(N-n)!n!}$$

(3) The number of different ways in which n indistinguishable particles can be arranged in g distinguishable states with not more then one particle in each state is equal to

$$\frac{g!}{(g-n)!n!}$$

2.14 MAXWELL-BOLTZMANN DISTRIBUTION LAW

Consider a system that contains a large number of particles that are identical and distinguishable. The identical particles refer to the particles having the same structure. The distinction between the particles is due to their energy states at a given instant. As shown in Fig. 2.8, 4 particles are in energy state E_1, 2 in state E_2, zero in state E_3 and so on. It is assumed that all the energy states are accessible to each particle. Consequently, it can be assumed that the probability of any particular partition is proportional to the number of different ways in which the particles can be distributed in the existing available energy states so as to produce the desired partition.

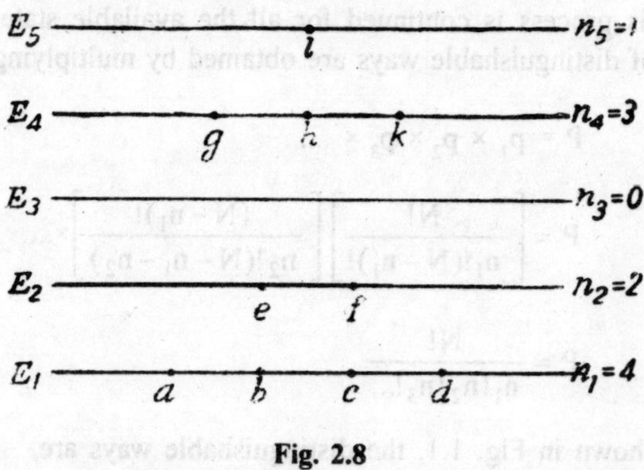

Fig. 2.8

To obtain the required distribution as shown in Fig. 2.8, the first particle a in state E_1 can be selected in N ways. The second particle b in state E_1 can be selected in (N – 1) ways and so on. Therefore, the total number of ways in which the first four particles in state E_1 can be selected is given by,

$$N(N-1)(N-2)(N-3) = \frac{N!}{(N-4)!}$$

Moreover, the four particles in state E_1 can be arranged in 4! different orders. For example, *abcd, bcda, cdab* and so on. There are 24 ways. But, it is immaterial for these particles to be arranged in any particular order in state E_1, because they are identical. Thus, the total number of *distinguishable* different ways are,

$$\frac{N!}{4!(N-4)!}$$

In general, if the first state consists of n_1 particles, the distinguishable different ways, for arranging n_1 particles in state E_1, are,

$$p_1 = \frac{N!}{n_1!(N - n_1)!} \quad ...(i)$$

For the second state E_2, only $(N - n_1)$ particles are available and n_2 particles are in state E_2. The number of distinguishable different ways are,

$$p_2 = \frac{(N - n_1)!}{n_2!(N - n_1 - n_2)!} \quad ...(ii)$$

If this process is continued for all the available states, the total number of distinguishable ways are obtained by multiplying P_1, P_2, P_3 etc.

$$P = p_1 \times p_2 \times p_3 \times \text{......}$$

$$P = \left[\frac{N!}{n_1!(N - n_1)!}\right]\left[\frac{(N - n_1)!}{n_2!(N - n_1 - n_2)}\right] \times \text{....}$$

$$P = \frac{N!}{n_1!n_2!n_3!\text{...}} \quad ...(iii)$$

As shown in Fig. 1.1, the distinguishable ways are,

$$P = \frac{N!}{4!2!0!3!1!} \quad ...(iv)$$

Here 0! is equal to one.

It has been assumed so far that all the available states have the same probability of occupation by the particles. However, it may happen that the states have different intrinsic probabilities say g_i. For example, a particular energy state may be favourable with more different angular momentum states than the rest and hence it is more likely to be occupied. Taking into account this intrinsic probability factor, the value of P will be different.

If g_i is the probability of locating a particle in a certain energy state E_i, then the probability of locating 2 particles in the same state is $g_i \times g_i = g^2{}_i$. For n_i particles, the probability is $g_i n_i$. Hence the total probability for a given distribution is given by

$$P = \frac{N!g_1n_1g_2n_2g_3n_3\text{...}}{n_1!n_2!n_3!n_4!\text{...}} \quad ...(v)$$

Here n_1, n_2, n_3 etc., are the number of particles in states E_1, E_2, E_3, etc. and g_1, g_2, g_3 etc., are the intrinsic probabilities for states E_1, E_2, E_3 etc.

If all the particles are further assumed to be indistinguishable *i.e.*, particles in state E, and particles in state E_4 cannot be distinguished, then N! permutations among the particles themselves and occupying the different states result in the same distribution. The probability in this case is given by

$$P = \frac{1}{N!}\left[\frac{N! g_1 n_1 g_2 n_2 g_3 n_3 \ldots}{n_1! n_2! n_3! \ldots}\right]$$

$$\therefore \quad P = \frac{g_1 n_1 g_2 n_2 g_3 n_3 \ldots}{n_1! n_2! n_3! \ldots}$$

$$\therefore \quad P = \prod_{i}^{N} \frac{g_i n_i}{n_1!} \qquad \ldots(vi)$$

Here Π is the product sign (since probability is the product of such distributions.

The most probable distribution can be obtained by evaluating the maximum value of $\log_e$, P in equation (vi). This should also satisfy the two conditions that

$$\sum_i n_i = N \qquad \ldots(vii)$$

and $$\sum_i n_i \; E_i = U \qquad \ldots(vii)$$

According to Stirling's approximation

$$\log_e x! = x \log_e x - x$$

From equation (vi), applying Stirling's approximation, we get

$$\log_e P = \sum_i (n_i \log_e g_i - \log_e n_i!)$$

$$= \sum_i [n_i \log_e g_i - (n_i \log_e n_i - n_i)]$$

$$= \sum_i (n_i \log_e g_i - n_i \log_e n_i + n_i)$$

$$= \sum_i n_i - \sum_i (n_i \log_e n_i - n_i \log_e g_i)$$

$$\log_e P = N \sum_i n_i \log_e \frac{n_i}{g_i} \quad \text{...(ix)}$$

Differentiating equation (ix)

$$d(\log_e P) = \sum_i (dn_i) \log_e \frac{n_i}{g_i} - \sum_i n_i d\left(\log_e \frac{n_i}{g_i}\right)$$

$$= \sum_i dn_i \log_e \frac{n_i}{g_i} - \sum_i n_i \frac{dn_i}{n_i}$$

$$= \sum_i dn_i \log_e \frac{n_i}{g_i} - \sum_i dn_i$$

But $\sum_i dn_i = 0.$

$$\therefore \quad -d(\log_e P) = \sum_i \left[\log_e \left(\frac{n_i}{g_i}\right)\right] dn_i \quad \text{...(x)}$$

To obtain the maximum value of P,

$$d(\log_e P) = 0$$

$$\therefore \quad \sum_i \log_e \left(\frac{n_i}{g_i}\right) dn_i = 0 \quad \text{...(xi)}$$

But $\Sigma dn_i = 0$ ...(xii)

and $\Sigma E_i dn_i = 0$ ...(xiii)

Multiplying (xii) by α and (xiii) by β and adding to equation (xi), we get

$$\sum_i \left[\log_e \left(\frac{n_i}{g_i}\right) + \alpha + \beta E_i\right] dn_i = 0$$

The equilibrium distribution is possible if

$$\log_e \frac{n_i}{g_i} + \alpha + \beta \, Ei = 0 \quad \text{...(xiv)}$$

or $$\frac{n_i}{g_i} = e^{-\alpha - \beta Ei}$$

or $$n_i = g_i \, e^{-\alpha - \beta Ei} \quad \text{...(xv)}$$

This gives the maximum probability distribution and α and β are two parameters that depend upon the physical property of the system.

$$N = n_1 + n_2 + n_3 + \ldots$$

$$N = g_1 e^{-\alpha-\beta E1} + g_1 e^{-\alpha-\beta E2} + \ldots.$$

$$N = e^{-\alpha}[g_1 e^{-\beta E1} + g_2 e^{-\beta E2} + \ldots.]$$

$$N = e^{-\alpha}[\sum_i g_i e^{-\beta Ei}]$$

Take $[\sum_i g_i e^{-\beta Ei} = Z$ ...(xvi)

Here, Z is called the partition function.

$$N = e^{-\alpha}(Z) \qquad \text{....(xvii)}$$

$$e^{-\alpha} = \frac{N}{Z}$$

Substituting this value in equation (vii)

$$n_i = \left(\frac{N}{Z}\right)[g_i e^{-\beta Ei}] \qquad \text{....(xviii)}$$

This equation refers to Maxwell-Boltzmann Distribution Law.

2.15 MAXWELL-BOLTZMANN DISTRIBUTION IN TERMS OF TEMPERATURE

The total energy of an isolated system is given by

$$U = \sum_i n_i E_i$$

$$U = n_1 E_1 + n_2 E_2 + n_3 E_3 + \ldots$$

$$U = (g_1 e^{-\alpha-\beta E1}) E_1 + (g_2 e^{-\alpha-\beta E2}) E_2 + \ldots$$

$$U = e^{-\alpha}[g_1 E_1 e^{-\beta E1} + g_2 E_2 e^{-\beta E2} + \ldots]$$

But $e^{-\alpha} = \frac{N}{Z}$

$$\therefore \quad U = \frac{N}{Z}[g_1 E_1 e^{-\beta E1} + g_2 E_2 e^{-\beta E2} + \ldots]$$

$$U = \frac{N}{Z}\sum_i g_i E_i e^{-\beta Ei} \qquad \text{...(i)}$$

Here $Z = \sum_i g_i e^{-\beta Ei}$ ...(i)

$$\frac{dZ}{d\beta} = \frac{d}{d\beta}\sum_i g_i e^{-\beta Ei} = -\sum_i g_i E_i e^{-\beta Ei}$$

$$\therefore \qquad \sum_i g_i E_i e^{-\beta Ei} = \frac{d}{d\beta} \sum_i g_i E_i e^{-\beta Ei}$$

Substituting this value in equation (i)

$$U = -\frac{N}{Z} \frac{d}{d\beta} \sum_i g_i e^{-\beta Ei}$$

$$U = -\frac{N}{Z} \frac{d}{d\beta}$$

$$U = -N \frac{d}{d\beta} [\log_e Z] \qquad \text{...(ii)}$$

The average energy of a particle is

$$E_{av} = -\frac{U}{N} = -\frac{d}{d\beta} (\log_e Z) \qquad \text{...(iii)}$$

This shows that for a given system, the total energy U, the partition function Z, and the average energy of the particle E_{av}, depend on the parameter β. Therefore, β may be taken to characterise the internal energy of the system and has the units per joule. If T is the temperature in degrees kelvin, then it is more convenient to represent parameter β as

$$\beta = \frac{1}{kT} \text{ or } KT = \frac{1}{\beta} \qquad \text{...(iv)}$$

Here kT has the unit of energy *i.e.*, joule and k is the Boltzmann's constant. Its units are J/K. The value of k is given by

$$k = 1.3805 \times 10^{-23} \text{ J/K}$$

Substituting the value of

$$\beta = \frac{1}{kT}$$

in all the equations, we get

$$Z = \sum_i g_i e^{-(Ei/kT)} \qquad \text{...(v)}$$

$$n_i = \frac{N}{Z} \sum_i g_i e^{-(Ei/kT)} \qquad \text{...(vi)}$$

This equation represents the Maxwell-Boltzmann distribution law in terms of the temperature of the system.

As $\qquad \beta = \frac{1}{kT}$

$$d\beta = -\frac{dT}{kT^2}$$

Substituting this value in equation (ii)

$$\therefore \quad U = (kNT^2)\frac{d}{dT}(\log_e Z) \qquad ...(vii)$$

Also, from equation (iii)

$$E_{av} = (kT^2)\frac{d}{dT}(\log_e Z) \qquad ...(viii)$$

Equation (viii) gives the relation between the average energy of the particle and its temperature under equilibrium position. Hence, the temperature of a system in statistical equilibrium is the physical quantity related to the average energy of the particle of the system.

2.16 MAXWELL-BOLTZMANN DISTRIBUTION AND IDEAL GAS

Most of the gases obey Maxwell-Boltzmann distribution over a wide range of temperature. Consider a gas having mono-atomic molecules. It is assumed that the gas is ideal and has only kinetic energy of translation. Intermolecular attraction is assumed to be absent. If the kinetic energy is not quantized but considered to possess continuous values of energy, then the partition function Z can be written as

$$Z = \int_0^\infty e^{-(E/kT)}\, g(E)dE$$

Here g_i is replaced by g (E) dE

$$\text{But} \quad g(E)dE = \left(\frac{4\pi V(2m^3)^{1/2}}{h^3}\right)E^{1/2}dE$$

Here V is the volume occupied by the gas and h is Planck's constant.

$$Z = \frac{4pV(2m^3)^{1/2}}{h^3}\int_0^\infty E^{1/2}e^{-(E/kT)}dE$$

$$Z = \frac{4\pi V(2m^3)^{1/2}}{h^3}\left[\frac{\sqrt{\pi(kT)^3}}{2}\right]$$

$$Z = \frac{V(2\pi mkT)^{3/2}}{h^3} \qquad ...(i)$$

Equation (i) represents the partition function for an ideal monoatomic gas in terms of the volume and temperature of the gas.

Taking logarithms on both sides

$$\log_e Z = \log_e \left[\frac{(2\pi m)^{3/2} V}{h^3}\right] + \log_e (kT)^{3/2}$$

$$\log_e Z = C + \frac{3}{2}\log_e kT \qquad \text{...(ii)}$$

Also $\quad E_{av} = (kT^2)\frac{d}{dT}(\log_e Z)$

$$E_{av} = (kT^2)\frac{d}{dT}\left[C + \frac{3}{2}\log_e kT\right]$$

$$= (kT^2)\frac{d}{dT}\left[C + \frac{3}{2}\log_e k + \frac{3}{2}\log_e T\right]$$

$$E_{av} = \frac{3}{2}kT \qquad \text{...(iii)}$$

The total energy,

$$U = NE_{av}$$

$$U = N\left(\frac{3}{2}kT\right)$$

$$U = \frac{3}{2}NkT. \qquad \text{...(iv)}$$

Equation (iv) shows that the internal energy of an ideal monoatomic gas depends only on its temperature. The same relation does not however, hold good for real gases. In the case of real gases the internal energy is partly potential and partly kinetic. The total energy depends on the volume of the gas.

2.17 QUANTUM STATISTICS

In Maxwell-Boltzmann distribution, it has been assumed that all the energy levels are accessible to all the particles of the system. However, there may be certain levels prohibited to a certain group of particles. Each energy state is associated with a certain available wave function.

The probability of a particular distribution is restricted by the available wave functions of any state. These restrictions are taken into account in quantum statistics. There are two types of quantum statistics

(i) Fermi-Dirac statistics

(ii) Bose-Einstein statistics.

In Fermi-Dirac statistics the particles are assumed to obey Pauli's exclusion principle. They are characterised by antisymmetric wave functions. The particles under, this category are called *fermions*. Protons neutrons and electrons are fermions.

In the case of Bose-Einstein statistics, it is assumed that the particles are not restricted by Pauli's exclusion principle and are characterised by symmetric wave functions. These particles are called *bosons*. It has been experimentally found that all particles having spin zero or integral multiple of 1 are bosons. Helium nuclei and mesons are bosons.

In both the kinds of quantum statistics, the particles are identical and indistinguishable. At high temperature and low pressure, all the three statistics give practically the same result.

2.18 FERMI-DIRAC DISTRIBUTION LAW

Consider a system consisting of a large number of particles. It is assumed:

(i) that the particles are *identical* and *indistinguishable*, and

(ii) that the particles obey exclusion principle. It means that no two particles can have the same dynamical state and the wave function of the whole system must be antisymmetric. The particle satisfying these conditions are called *fermions*. In general, all fundamental particles with spin are fermions.

In quantum statistics the intrinsic probability g_i is governed by the different quantum states relating to a given energy *i.e.*, the degeneracy of the energy state. Each quantum state corresponds to a particle wave function. The Wave functions are determined by each of the possible arrangements of quantum number corresponding to a given energy level. For particles with spin 1/2, in the absence of magnetic forces, each of the particle may be in the energy states with spin +1/2 or –1/2. Hence the intrinsic probability g_i in this case is 2. For motion in a central field, the energy of the particle is independent of the orientation of the orbital angular momentum Due to this, a degeneracy of $(2\, l + 1)$ is introduced and it is the value of g for that particular energy state. If the particles possess spin, the total degeneracy = $2\,(2l + 1)$. As no two particles can be in the same energy state having the same quantum number, the instrinsic probabilities g_i's give the maximum number of particles that can be accommodated in a particular energy level without violating the exclusion principle *i.e.*,

$$n_i \leq g_i$$

It means that the n_i value for a given distribution does not exceed the corresponding value of g_i.

Let n_i, be the number of particles for energy level E_i The first particle can be placed in any one of the available g_i states *i.e.*, this particle can be assigned to any of the gi sets of quantum numbers. Thus, the first particle can be distributed in g_i different ways. Similarly, the second particle can be arranged in $(g_i - 1)$ different ways and the process continues.

Thus, the total number of different ways of arranging n_i particles among the available g_i states with energy level E_i is

$$= g_i(g_i - 1)(g_i - 2) \;......[g_i - (n_i - 1)]$$

$$= \frac{g_i!}{(g_i - n_i)!} \qquad ...(i)$$

Further, if the particles are taken to be indistinguishable, it will not be possible to detect any difference when n_i particles are reshuffled into different states occupied by them in the energy level E_i. Therefore, the total number of different and distinguishable ways is,

$$= \frac{g_i!}{n_i!(g_i - n_i)!} \qquad ...(ii)$$

Therefore, the total number of different and distinguishable ways of getting the distribution n_1, n_2, n_3, etc., among the various energy levels, E_1, E_2, E_3...,etc.. can be obtained by multiplying the various factors.

$$\therefore \quad P = \frac{g_1!}{n_1!(g_1 - n_1)!} \cdot \frac{g_{2i}!}{n_2!(g_2 - n_2)!} \;....$$

$$P = \prod_i \frac{g_i!}{n_i!(g_i - n_i)!} \qquad ...(iii)$$

The most probable distribution can by obtained by evaluating the maximum value of $\log_e$, P in equation (iii).

This should also satisfy the condition that

$$\sum_i n_i = N$$

and $$\sum_i n_i E_i = U.$$

According to Stirling's approximation

$$\log_e x! = x \log_e x - x$$

From equation (iii), applying Stirling's approximation,

$$\log_e P = \sum_i [(g_i \log_e g_i - g_1) - (n_i \log_e n_i - n_i)$$

$$- [(g_i - n_1) \log_e (g_i - n_1) - (g_i - n_1)]$$

$$\log_e P = \sum_i [g_i \log_e g_i - n_1 - \log_e n_i - (g_i - n_i)$$

$$\log_e (g_i - n_1)] \quad ....(iv)$$

Differentiating equation (iv)

$$-d (\log_e P) = \sum_i [\log_e n_i - \log_e (g_i - n_i)]dn_i$$

To obtain the maximum value of P,

$$d(\log_e P) = 0$$

$$\therefore \quad \sum_i [\log_e n_i - \log_e (g_i - n_i)]dn_i = 0 \quad ...(v)$$

But $$\sum_i dn_i = 0 \quad ...(vi)$$

and $$\sum_i E_i dn_i = 0 \quad ...(vii)$$

Multiplying (vi) by α and (vii) by β and adding to equation (v), we get

$$\sum_i [\log_e n_i - \log_e (g_i - n_1) + \alpha + \beta E_i]dn_i = 0$$

The equilibrium distribution is possible if

$$\therefore \quad \log_e n_i - \log_e (g_i - n_1) + \alpha + \beta E_i = 0 \quad ...(viii)$$

$$\log_e\left(\frac{n_i}{g_i - n_i}\right) = -\alpha + \beta E_i$$

$$\frac{n_i}{g_i - n_i} = e^{-\alpha - \beta Ei}$$

$$\frac{g_i - n_i}{n_i} = e^{\alpha + \beta Ei}$$

$$\frac{g_i}{n_i} - 1 = e^{\alpha + \beta Ei}$$

$$\frac{g_i}{n_i} = (e^{\alpha + \beta Ei}) + 1$$

$$n_i = \left(\frac{g_i}{e^{\alpha+\beta E_i}}\right) + 1 \qquad \text{...(ix)}$$

Equation (ix) represents the Fermi-Dirac distribution law. The parameter has the same role as in the case of Maxwell-Boltzmann distribution law *i.e.*, for the system consisting of fermions in statistical equilibrium.

$$\beta = \frac{1}{kT}$$

$$\therefore \quad n_i = \frac{g_i}{[e^{(\alpha+E_i/kT)}]+1}$$

In most cases the value of a is negative and is taken to be equal to

$$\frac{-E_F}{kT}$$

$$n_i = \frac{g_i}{\left[e^{(E_i-E_F)/kT)}\right]+1}$$

The value of E_F is positive and is independent of temperature.

For T = 0, all the energy states are fully occupied and $n_i = g_i$ All the states with $E > E_F$, are empty *i.e.*, $n_i = 0$.

$$\left[e^{(E_i-E_F)/kT)}\right] = \begin{cases} 0 \text{ for } E_i - E_F < 0 \\ \infty \text{ for } E_i - E_F > 0 \end{cases}$$

Limit T $\rightarrow$ 0

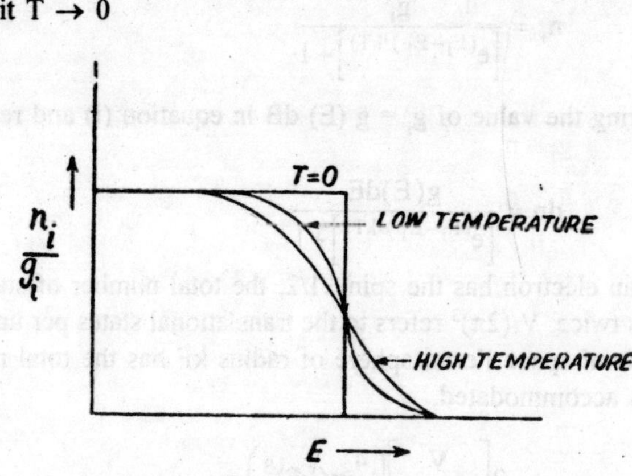

Fig. 2.9

In the case of Fermi-Dirac statistics, the accumulation of particles at the ground level is not allowed and at temperature T = 0, the particles occupy the lowest energy levels upto E_F. Here, the energy E_F gives the indication of the maximum energy of the fermions in the system. By is also called Fermi energy. For higher temperatures, the particles, occupy higher energy states greater than E_F. The curves (Fig. 2.9) indicate that only those fermions with energies close to E_F can move into unoccupied higher energy states. If $k\theta_F = E_F$, the temperature θ_F is called the Fermi temperature.

2.19 ELECTRON GAS

Electrons in a metal belong to a most characteristic system of fermions because electrons obey the exclusion principle. For electrons in a metal, the energy levels are grouped in bands. Practically at all temperatures, the lower level energy bands are filled with electrons. The upper level energy bands are only partially filled with electrons. The distribution of electrons is to be considered only in the upper bands called the conduction band. The zero energy level is taken at the lowest level of the conduction band. It is also assumed that the electrons have free movement within the conductor, provided the energy associated with the electrons is of the order of upper level energy bands. As the energy of the electron in the conduction band is continuous, the term g_i is replaced by g(E) dE. Here dn electrons have energy in the range E and E + dE. According to Fermi-Dirac distribution law

$$n_i = \frac{g_i}{\left[e^{(E_i - E_F)/kT}\right] + 1} \qquad \text{...(i)}$$

Substituting the value of g_i = g (E) dB in equation (i) and replacing n_i by dn

$$dn = \frac{g(E)dE}{\left[e^{(E_i - E_F)/kT}\right] + 1}$$

As an electron has the spin ±1/2, the total number of states in the sphere is twice. $V/(2\pi)^3$ refers to the translational states per unit volume in the Fermi space Fermi sphere of radius kF has the total number of particles accommodated,

$$n = \frac{2\left[\frac{V}{(2\pi)^3}\right]\left(\frac{4}{3}\pi(kF)^3\right)}{\left[e^{(E_i - E_F)/kT}\right] + 1}$$

Here $$E = \frac{p^2}{2m} = \frac{\left(\frac{h}{2\pi}\right)^2 (kF)^2}{2m}$$

$$\therefore \quad kF = \frac{p}{h/2\pi} = \frac{2\pi p}{h}$$

$$\therefore \quad n = \frac{2\left[\frac{V}{(2\pi)^3}\right]\left(\frac{4}{3}\pi\left(\frac{2\pi p}{h}\right)^3\right)}{\left[e^{(E_i - E_F)/kT}\right] + 1}$$

But $p = (2mE)^{1/2}$

$$\therefore \quad n = \frac{2\left[\frac{V}{(2\pi)^3}\right]\left[\left(\frac{4}{3}\pi\left(\frac{2\pi p}{h}\right)^3 (2mE)^{3/2}\right)\right]}{\left[e^{(E_i - E_F)/kT}\right] + 1}$$

$$n = \frac{\left(\frac{8\pi V}{(3h^3}\right)(2m)((2m)^{1/2} E^{3/2}}{\left[e^{(E_i - E_F)/kT}\right] + 1}$$

Differentiating

$$dn = \frac{\left(\frac{8\pi V}{(3h^3}\right)(2m)((2m)^{1/2} \times \frac{3}{2} E^{1/2} dE}{\left[e^{(E_i - E_F)/kT)}\right] + 1}$$

$$\frac{dn}{dE} = \frac{\left[\frac{8\pi V(2m^3)^{1/2}}{(3h^3}\right] E^{1/2}}{\left[e^{(E_i - E_F)/kT)}\right] + 1} \qquad ...(ii)$$

Equation (ii) represents the energy distribution for free electrons. This is also called Fermi-Dirac formula of free fermions.

At $T = 0K$,

$$p = (2m\ E_F)^{1/2} \qquad ...(iii)$$

and $N = \left(\frac{2V}{h^3}\right)\left(\frac{4}{3}\pi p^3\right)$

or $$P = \left(\frac{3Nh^3}{8\pi V}\right)^{1/3} \qquad ...(iv)$$

Equating (iii) and (iv)

$$(2mE_F)^{1/2} = \left(\frac{3Nh^3}{8\pi V}\right)^{1/3}$$

Squaring

$$2mE_F)^{1/2} = \left(\frac{3Nh^3}{8\pi V}\right)^{2/3}$$

or $$E_F = \frac{h^2}{8m}\left(\frac{3N}{\pi V}\right)^{2/3} \qquad ...(v)$$

Knowing the value of N/V *i.e.*, the number of free electrons per unit volume, the Fermi energy for electrons in a metal can be obtained.

For silver, the value $\frac{N}{V} = 5.86 \times 10^{28}$ electrons/m³

$$\therefore \qquad E_F = \left(\frac{(6.624 \times 10^{-34})^2}{8 \times 9 \times 10^{-31}}\right)\left(\frac{3 \times 5.86 \times 10^{28}}{3.14}\right)^{2/3}$$

$$\mathbf{E_F = 9 \times 10^{-19} J.}$$

But $1 eV = 1.6 \times 10^{-19} J$

$$\therefore \quad E_F = \frac{9 \times 10^{-19}}{1.6 \times 10^{-19}}$$

$$\mathbf{E_F = 5.625\ eV.}$$

It means that the maximum kinetic energy of free electrons in silver at absolute zero temperature is 5.6 eV.

The Fermi temperature θ_F is given by

$$\theta_F = \frac{E_F}{k}$$

As E_F practically independent of temperature, the value θ_F is fixed for a given metal. The values of θ_F and E_F are given in the following table.

Metals	**E_F (electron volts)**	**θ_F(K)**
Potassium	2.14	2.4×10^4
Sodium	312	3.7×10^4
Lithium	4.72	6.5×10^4
Silver	5.51	6.4×10^4
Gold	5.64	6.4×10^4
Copper	7.04	8.2×10^4

2.20 BOSE-EINSTEIN DISTRIBUTION LAW

Bose-Einstein distribution is applied to systems composed of identical and indistinguishable particles that are not restricted by the exclusion principle. In such systems, there is no limit to the number of particles occupying a particular quantum state. These particles are called *bosons.* Their spin is zero or 1. Mesons and helium nuclei are examples of bosons.

In the case of Bose-Einstein statistics the values of g_i refer to degeneracy of each energy level. Suppose, that n_i particles are arranged in a row and distributed among g_i quantum states with $(g_i - 1)$ partitions in between. The total number of possible arrangements of particles and partitions is equal to the total number of permutations of $(n_i + g_i - 1)$ objects in a row. Therefore the total possible ways of arranging n_i particles with $g_i - 1$ partitions

$$= (n_i + g_i - 1)!$$

As the particles are identical and indistinguishable the possible number of distinct arrangements

$$= \frac{(n_i + g_i - 1)!}{n_i!(g_i - 1)!}$$

The total number of distinguishable and distinct ways of arranging N particles in all the available energy states is given by

$$P = \frac{(n_1 + g_1 - 1)!}{n_1!(g_1 - 1)!} \times \frac{(n_2 + g_2 - 1)!}{n_2!(g_2 - 1)!} \times \ldots$$

$$P = \prod_i \frac{(n_i + g_i - 1)!}{n_i!(g_i - 1)!} \quad \text{...(i)}$$

The most probable distribution can be obtained by finding the maximum value of $\log_e$, P.

According to Stirling's approximation

$$\log_e x! = x \log_e x - x$$

Also $\sum_i n_i = N$ ...(ii)

and $\sum_i n_i E_i = U$ ...(iii)

From equation (i), applying Stirling's approximation

$$\log_e P = \sum_i [\log_e (n_i + g_i - 1) - \log_e n_i ! \log_e (g_i - 1)!]$$

$$\log_e P = \sum_i [(n_i + g_i - 1) \log_e (n_i + g_i - 1) - (n_i + g_i - 1)$$

$$- (n_i \log_e n_i - n_i) - (g_i - 1) \log_e (g_i - 1) - (g_i - 1)]$$

$$\log_e P = \sum_i [(n_i + g_i - 1) \log_e (n_i + g_i - 1)$$

$$- n_i \log_e - n_i - (g_i - 1) \log_e (g_i - 1)] \quad \text{...(iv)}$$

The maximum value of P is obtained by taking

$$d (\log_e P) = 0.$$

Differentiating equation P.

$$d (\log_e P) = \sum_i [\log_e (n_i + g_i - 1)\, dn_i - \log_e n_i - dn_i] = 0$$

$$\therefore \quad -d(\log_e P) = \sum_i [-\log_e (n_i + g_i - 1)\, dn_i + \log_e n_i dn_i] = 0$$

or

$$\sum_i [-\log_e (n_i + g_i - 1) + \log_e n_i] dn_i = 0 \quad \text{...(v)}$$

As the total number of particles and total energy are constants, we have,

$$\sum_i dn_i = 0 \quad \text{...(vi)}$$

$$\sum_i E_i dn_i = 0 \quad \text{...(vii)}$$

Multiplying (vi) by α and equation (vii) by β and adding to equation (v), we get

$$\sum_i [-\log_e (n_i + g_i - 1) + \log_e n_i + \alpha + \beta E_i] dn_i = 0$$

$$\therefore \quad -\log_e (n_i + g_i - 1) + \log_e n_i + \alpha + \beta E_i = 0$$

Taking $n_i + g_i$ very large as compared to 1, the quantity 1 can be neglected.

$$\therefore \quad -\log_e (n_i + g_i) + \log_e n_i + \alpha + \beta E_i = 0$$

$$\log_e \left(\frac{n_i}{ng_i + g_i} \right) = -\alpha - \beta E_i$$

$$\frac{n_i}{n_i + g_i} = e^{-\alpha - \beta Ei}$$

$$\frac{n_i + g_i}{n_i} = e^{\alpha + \beta Ei}$$

$$1 + \frac{g_i}{n_i} = e^{\alpha + \beta Ei}$$

$$\frac{g_i}{n_i} = (e^{\alpha + \beta Ei}) - 1$$

$$n_i = \left(\frac{g_i}{e^{\alpha + \beta E_i}} \right) - 1 \quad \text{...(viii)}$$

Equation (ix) represents the Bose-Einstein distribution law.

Taking $\quad \beta = \dfrac{1}{kT}$

$$\therefore \quad n_i = \frac{g_i}{[e^{(\alpha + E_i/kT)}] - 1}$$

The value of the constant a is governed by the equation $\Sigma n_i = N$. As n_i cannot be negative, α must always have a positive value. The distribution of the particles for different energy levels is shown in Fig. 2.10 for low and high temperatures.

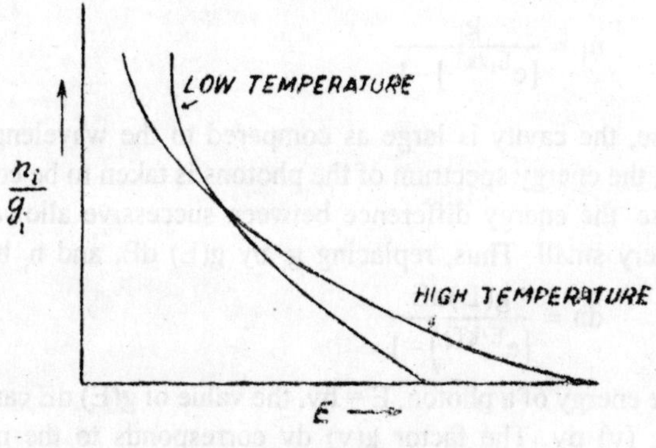

Fig. 2.10

The Bose-Einstein distribution has more emphasis on the lower energy levels.

2.21 PHOTON GAS

The interaction of electromagnetic radiations with matter, led to the idea that electromagnetic radiations are composed of discrete energy particles called photons. Each photon has an energy $h\nu$ and momentum h/λ. Here ν is the frequency and λ is the wavelength of the radiations. The electromagnetic radiations trapped in a cavity and in thermal equilibrium with the walls of the cavity are termed as black body radiations. In the equilibrium condition, the black body radiations can be considered as the *photon gas*. It is assumed that the photons do not interact among themselves. The photons interact only with-the atoms of the walls of the cavity. It is further assumed that the photons are indistinguishable and many photons can have the same energy. Photons are taken as bosons and they obey Bose-Einstein statistics.

As the photons can either be emitted or absorbed by the atoms of the walls of the cavity, the number of photons is not constant *i.e.*, the condition $\sum_i dn_i = 0$ is no longer valid.

Due to this reason the value of α is equal to zero in the Bose-Einstein distribution law.

$$n_i = \frac{g_i}{[e^{(\alpha + E_i/kT)}] - 1} \quad ...(i)$$

Here $\alpha = 0$

$$\therefore \quad n_i = \frac{g_i}{[e^{E_i/kT}] - 1} \quad ...(ii)$$

In case, the cavity is large as compared to the wavelength of the radiations, the energy spectrum of the photons is taken to be continuous. In this case, the energy difference between successive allowed energy value is very small. Thus, replacing g_i by g(E) dE, and n_i by dn

$$dn = \frac{g(E)dE}{\left(e^{E/kT)}\right) - 1} \quad ...(iii)$$

As the energy of a photon, E = hν, the value of g(E) dE can be taken equal to g (ν) dν. The factor g(ν) dν corresponds to the number of oscillatory modes in the frequency range dν and relating to the energy range dE.

The number of states in a black body radiation in the frequency range ν and ν + dν can be obtained by calculating the spherical volume bound by the spheres of radii $\frac{h(\nu + d\nu)}{c}$ and $\frac{h\nu}{c}$

The volume of the spherical shell

$$= \frac{4}{3}\pi \frac{h^3}{c^3}(\nu + d\nu)^3 - \frac{4}{3}\pi \frac{h^3}{c^3}\nu^3$$

$$= \frac{4}{3}\pi \frac{h^3}{c^3}[\nu^3 + 3\nu^2 d\nu + ... - \nu^3]$$

$$= \frac{4}{3}\pi \frac{h^3}{c^3} \times 3\nu^2 d\nu$$

$$= 4\pi \frac{h^3}{c^3} \nu^2 d\nu \quad ...(iv)$$

The phase space has volume $V = h^3$ and there are two states of polarization for the radiation. The number of states in the black body radiation in the frequency range ν and ν + dν is given by

$$g(\nu)d\nu = \frac{8\pi\nu}{c^3}\nu^2 d\nu \quad ...(v)$$

$$dn = \left(\frac{8\pi V\nu^2 d\nu}{c^3}\right)\frac{1}{\left(e^{h\nu/kT}\right)-1} \quad ...(vi)$$

For dn photons in the frequency range ν + dν and ν, the energy is equal to (hν) dn and the energy per unit volume

$$= \frac{(h\nu)dn}{\nu}$$

Therefore, the energy density distribution for black body radiation is given by

$$E(\nu) = \frac{(h\nu)dn}{V.d\nu} \quad ...(vii)$$

Substituting the value of dn from equation (vi), we get

$$E(\nu) = \frac{8\pi h\nu^2 d\nu . h\nu}{c^3 V.d\nu\left[\left(e^{h\nu/kT}\right)-1\right]}$$

$$E(\nu) = \left(\frac{8\pi h\nu^3}{c^3}\right)\left[\frac{1}{\left(e^{h\nu/kT}\right)-1}\right]^{d\nu} \quad ...(viii)$$

Equation (viii) represents Planck's radiation law for black body radiations.

From equation (viii)

$$E(\nu)d\nu = \frac{8\pi h\nu^3}{c^3}\left[\frac{1}{\left(e^{h\nu/kE}\right)-1}\right]^{d\nu}$$

But $\nu = \frac{c}{\lambda}$

or $d\nu = -\frac{c}{\lambda^2}d\lambda$

Neglecting the negative sign, we get

$$\therefore \quad E(\lambda)d\lambda = \frac{8\pi h\nu^3}{c^3} \times \frac{c}{\lambda^2}\left[\frac{1}{\left(e^{h\nu/kT}\right)-1}\right]^{d\lambda}$$

or $$E(\lambda) = \frac{8\pi hc}{\lambda^5} \times \frac{c}{\lambda^2}\left[\frac{1}{\left(e^{h\nu/kT}\right)-1}\right] \qquad ...(ix)$$

This gives the energy density for wavelength λ in the spectrum of the black body. Both the equations (viii) and (ix) represent Planck's radiation law for black body radiation.

SOLVED EXAMPLES

Example 1:

For N = 400, find the ratios of the probabilities for (300, 100) and (200, 200) distribution in two identical boxes.

Solution:

We have

$$P(300, 100) = \frac{400!}{300!\,100!}\,\frac{1}{2^{400}}$$

$$P(200, 200) = \frac{400!}{200!\,200!}\,\frac{1}{2^{400}}$$

$$\frac{P(300, 100)}{P(200, 200)} = \frac{200!\,200}{300!\,100!} \cong 10^{-5}$$

Example 2:

Eight similar coins are tossed, find the probability of three heads and t tails.

Solution:

We know

$$P(3, 5) = \frac{{}^8C_3}{2^8} = \frac{8\times7\times6}{3\times2\times1\times2^8}$$

or $$P(3, 5) = \frac{7}{32}$$

Example 3:

A system consists of 6000 particles distributed in three energy states with equal spacing. The energy of the three states are $E_{I'} = 0$, $E_{2'} = x$

and E_3 = 2x. All the three slates have the same intrinsic probability g. At a certain instant there are 3000 particles in lower level, 2500 in the middle level and 500 in the upper level. Compare the relative probabilities with the distribution obtained, by the transfer of one particle from the middle to the lower level and one particle from the middle to the upper level and the original distribution.

Solution:

Let P_1 and P_2 be the probabilities in the two cases

In the first case,

$$N = 6000,$$

$$n_1 = 3000,$$

$$n_2 = 2500,$$

$$n_3 = 500.$$

$$P_1 = \frac{g^N}{n_1!n_2!n_3!}$$

$$P_1 = \frac{g^{6000}}{3000!2500!500!} \quad ...(i)$$

In the second case

$$N = 6000,$$

$$n_1 = 3001,$$

$$n_2 = 2498,$$

$$n_3 = 501.$$

$$P_1 = \frac{g^N}{n_1!n_2!n_3!}$$

$$P_1 = \frac{g^{6000}}{3001!2498!501!}$$

Dividing (ii) by (i)

$$\frac{P_2}{P_1} = \frac{3000!2500!500!}{3001!2498!501!}$$

$$\frac{P_2}{P_1} = \frac{2500\times2499!}{3001\times501!}$$

$$\frac{P_2}{P_1} = 4.157$$

$$\approx 4.2.$$

It means the transfer of one particle from the middle to the upper and one particle from the middle to the lower state has changed the probability by a factor 4.2. This shows that both these distributions are not near the equilibrium state.

EXERCISES

1. Derive Planck's radiation law for the black body radiation.
2. Establish the distribution law of Maxwell and Boltzmann.
3. Discuss the free electron gas model for metals. Obtain the distribution law for the electron gas in a metal at absolute zero.
4. Distinguish between Classical Statistics, Fermi-Dirac Statistics and Bose-Einstein Statistics.
6. Obtain an expression for Bose-Einstein distribution law.
7. What is the principle of equal a priori probability (Write statistical postulates.
7. Explain the statistical significance of entropy and prove that

   $S = k \log W.$
8. What are the fundamental postulates of statistical mechanics? Obtain an expression for the number of states for a macroscopic system.
9. What is Gibbs ensemble? Differentiate between different types of ensembles.
10. Calculate the number of quantum states available to a proton inside a nucleus ($r = 10^{-14}$m) whose momentum does not exceed 10^{-19} kg m.s^{-1}.
11. An oven is at 600 K. Calculate the probability for 3 photons to be in a quantum state of energy 0.2 eV. Assume chemical potential of photon to be zero.
12. Show that as the molecules become numerous the deviation from most probable distribution is very rare.

13. How will you use statistical mechanics in deriving a relation for magnetisation in terms of temperature and external magnetic field B.
14. Comment "Uncertainty principle leads to statistical results."
15. Write short notes on: Entropy and probability, canonical ensemble, phase-space, macro and microstates, distribution of N molecules in m cells, degeneracy, quantum states.
16. Distinguish between classical and quantum statistics.
17. Explain what you understand by the term statistical equilibrium?
18. Discuss Maxwell-Boltzmann distribution law and cerive the relation.
19. On the basis of Maxwell-Boltzmann statistics show that the temperature of a system in statistical equilibrium is related to average energy of the particles of the system.
20. Applying Maxwell-Boltzmann distribution law, show that the internal energy of an ideal mono-atomic gas depends only on its temperature.
21. Write short notes on:
    (i) Statistical equilibrium
    (ii) Electron gas
    (iii) Photon gas
    (iv) Phase space
    (v) Quantum statistics
    (vi) Fermi temperature
    (vii) Fermi Energy
    (viii) Planck's radiation law
    (ix) Statistical thermodynamics
    (x) Maxwell Boltzmann distribution law
    (xi) Fermi-Dirac distribution law
    (xii) Bose-Einstein distribution law
22. A system consists of 8000 particles distributed in three energy states with equal spacing. The energy of the three states are

$E_1 = 0, E_2 = x. E_3 = 2x$. All the three states have the same intrinsic probability g. At a certain instant, there are 4000 particles in the lower level, 3000 particles in the middle level and 1000 particles in the upper level. Compare the relative probabilities with the distribution obtained by the transfer of one particle from middle to the lower level and one particle from the middle to the upper level and the original distribution.

23. A six faced dice is such that the probability for the number 4 is 1/5. Find the probability that 4 appears 4 times in 10 throws. Find also the ratio of this probability to the same complexion if the dice were normal.

24. You have 400 coins, out of them 20% are prone to 0.6 times head per throw. Find the probability of complexion 200 heads and 200 tails. Is it different from the maximum probability? Why?

25. If N = 300, find the ratio of the distribution 10% off from maximum probable, to the most probable one.

3

STEFAN'S LAW AND THERMAL CONDUCTORS

3.1 INTRODUCTION OF STEFAN'S LAW

The experimental study of the rate of emission of heat energy by a hot body by Tyndall helped Stefan (in 1879) to enunciate the law called Stefan's law in 1884, Boltzmann gave a theoretical proof of Stefan's Law on the basis of thermodynamics. Therefore, this law is also called Stefan-Boltzmann Law.

According to this law, *the rate of emission of radiant energy by unit area of a perfectly black body it directly proportional to the fourth power of its absolute temperature*

$$R \propto T^4$$

or $$R = \sigma T^4, \quad \text{...(i)}$$

where a is called Stefan's constant. If the body is not perfectly black and its emissivity or relative emittance is e, then

$$R = e\sigma T^4. \quad \text{...(ii)}$$

Hence e varies between zero and one, depending on the nature of the surface. For a perfectly black body e = 1. The law is not only true for emission but also for absorption of radiant energy. The hot body will continue emitting heat upto zero degree absolute temperature, if the temperature of the surroundings is zero degree absolute. But, in actual practice, the hot body is surrounded by a wall at some lower temperature. Due to this the body is continuously emitting and absorbing the heat radiations. When the body has the same temperature as that of the surroundings, the rate of emission and absorption are equal.

Hence, if a perfectly black body at temperature T_1 is surrounded by a wall (surroundings) at a temperature T_2 the net rate of loss (or gain) of heat energy per unit area of the surface is given by,

$$R \propto (T_1^4 - T_2^4)$$

$$R = \sigma(T_1^4 - T_2^4) \quad \text{...(iii)}$$

If the body has an emissivity e,

then $$R = e\sigma(T_1^4 - T_2^4) \quad \text{...(iv)}$$

3.2 MATHEMATICAL DERIVATION OF STEFAN'S LAW

The fact that blade body radiations exert pressure similar to a gas, helps in applying thermodynamics to heat radiations.

Let ψ be the energy density of radiations inside a uniform temperature enclosure at temperature T. P is the pressure and V is the volume.

Applying the first law of thermodynamics

$$\delta H = dU + P.dV \quad \text{...(i)}$$

Applying thermodynamical relation

$$\left(\frac{\partial H}{\partial V}\right)_r = \left(\frac{\partial P}{\partial T}\right)_v \quad \text{...(ii)}$$

$$\left(\frac{\partial U + P\partial V}{\partial V}\right)_r = T\left(\frac{\partial P}{\partial T}\right)_v$$

$$\left(\frac{\partial U}{\partial V}\right)_r = T\left(\frac{\partial P}{\partial T}\right)_v = P \quad \text{...(iii)}$$

Now $$U = V\psi$$

and $$P = \frac{\psi}{3}$$

or $$\left[\frac{\partial U}{\partial V}\right]_r = \psi$$

Here ψ is a function of temperature alone.

Substituting these values in equation (iii),

$$\psi = \frac{T}{3}\frac{d\psi}{dT} - \frac{\psi}{3}$$

$$\frac{4\psi}{3} = \frac{T}{3}\frac{d\psi}{dT}$$

or $$\frac{d\psi}{\psi} = 4\frac{dT}{T}$$

Integrating, $\log \psi = 4 \log T + \text{constant}$

or $$\psi = KT^4 \quad ...(iv)$$

Here K is a constant

Also the total rate of emission per unit area of a black body is proportional to the energy density.

$$\therefore \quad R \propto \psi \propto T^4$$

$$\therefore \quad R = \sigma T^4 \quad ...(v)$$

where σ is Stefan's constant.

The value of Stefan's constant in C.G.S. system is 5.672×10^{-5} C.G.S. units and in M.K.S. system it is 5.672×10^{-8} M.K.S. units.

3.3 DERIVATION OF NEWTON'S LAW OF COOLING FROM STEFAN'S LAW

Stefan's law is applicable for all temperatures of a hot body. But Newton's Law is applicable when the difference of temperature between the hot body and the surrounding is small. Consider a hot body at a temperature T_1 placed in a uniform temperature enclosure at T_1. According to Stefan's law,

$$R = e\sigma(T_1^4 - T_1^4) \quad ...(i)$$

Here e is the emissivity of the surface of the hot body

$$R = e\sigma(T_1 - T_2)(T_1^3 + T_1^2 T_2 + T_1 T_2^2 + T_2^3)$$

As $(T_1 - T_2)$ is small, T_1 can be taken approximately equal to T_2.

Then, $$R = e\sigma(T_1 - T_2)(T_2^3 + T_2^3 + T_2^3 + T_2^3)$$

$$R = 4e\sigma T_2^3((T_1 - T_2)$$

Taking $4e\sigma T_2^3 = k$

$$R = k(T_1 - T_2)$$

or $$R \propto (T_1 - T_2). \quad \text{...(ii)}$$

This equation represents Newton's law of cooling and is true when the difference of temperature is small.

3.4 EXPERIMENTAL VERIFICATION OF STEFAN'S LAW

In 1897, Lummer and Pringsheim experimentally verified Stefan's law over a wide range of temperature (100°C to 1,300°C).

The apparatus consisted of a black body C. For temperatures between 200°C and 600°C, a hollow copper sphere coated inside with platinum black was used. The fused nitrates of sodium and potassium having a melting point of 219°C were used as the bath surrounding the black body. For temperature between 900°C and 1,300°C, an iron cylinder coated inside with platinum black was used as a black body and it was enclosed in a double walled gas furnace. A thermocouple T was used as a thermometer. A bolometer B was used to measure the intensity of the emitted heat radiations. S_1, S_2 and S_3 were the water-cooling shutters (Fig. 3.1).

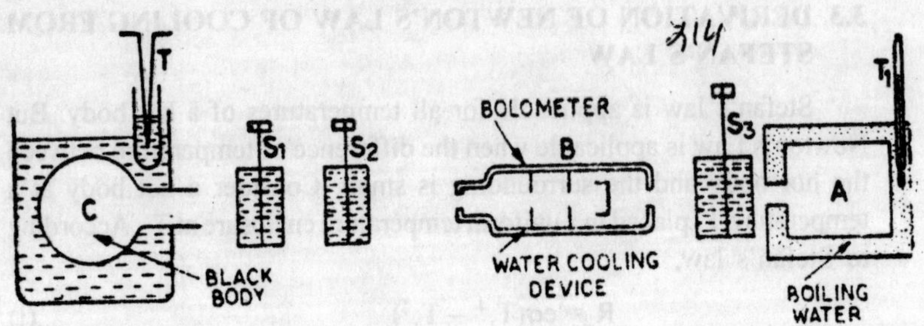

Fig. 3.1

Another black body A at 100°C was used to standardize the bolometer. The double walled vessel of the black body A contained boiling water at 100°C. The bolometer B was allowed to face the opening of the black body A and the shutter S_3 raised. The deflections in the galvanometer of the bolometer at various distances were noted and it was found that the deflection was inversely proportional to the square of the distance between the bolometer and the opening of the black body A. Thus, the deflection in the galvanometer was proportional to the intensity of heat radiations.

The shutter S_3 was closed and the bolometer B was allowed to face the opening of the black body C. The shutters of S_1 and S_2 were raised. The bath surrounding the black body was maintained at a constant temperature and the maximum deflection produced in the galvanometer of the bolometer was noted. Thus at various constant temperatures of the black body, corresponding to constant deflections (in the galvanometer of the bolometer) were observed. Then the data was reduced to a common arbitrary unit in terms of the total radiations from the black body A at 100°C.

Let θ be the deflection in the galvanometer, T_1 the temperature of the black body and T_2, the temperature at the entrance of the bolometer. It was found that

$$\theta \propto (T_1^4 - T_2^4)$$

But $$\theta \propto R$$

$\therefore$ $$\theta \propto (T_1^4 - T_2^4)$$

This verifies Stefan's law.

Recently Coblentz has verified Stefan's law more accurately. He took an electrically heated black body whose temperature was measured by an accurate thermocouple. An absolute bolometer was used to measure the amount of heat radiations emitted by the black body. He was able to show the correctness of Stefan's law experimentally up to 1,600°C.

3.5 DETERMINATION OF STEFAN'S CONSTANT (LABORATORY METHOD)

The laboratory apparatus used to determine the Stefan's constant is shown in Fig. 3.2. A hollow hemispherical metallic vessel A is enclosed in a wooden box W. The inner surface of A is coated with lamp black and the wooden box W is lined with tin plates. The whole apparatus is placed on a wooden base having a small hole at its centre. The vessel A is heated by passing steam inside the box and A acts as a black body radiator. The thermometers T T record the temperature of A.

A small silver disc B whose upper surface is coated with lamp black is placed at the central hole. The ebonite covering C is used to cover and uncover the disc B from the radiations of the enclosure. It can be arranged from outside with the help of the handle H. The disc B is connected to a thermocouple arrangement. One junction of the thermocouple is immersed in a tube containing oil. The tube is surrounded by a beaker containing water. A sensitive galvano- meter G is used in the circuit. The leads connected to the terminals of the galvanometer are immersed in cotton wool in the box F to avoid any distribution effect due to the difference of temperature in the leads. A rheostat Rh can be used in the circuit to obtain the deflection within the range. The actual experiment consists of two parts.

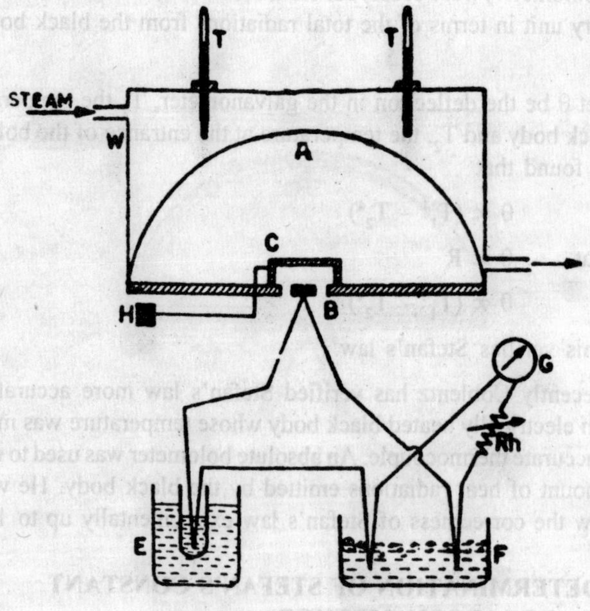

Fig. 3.2

The thermocouple is first standardized. Before passing steam into the chamber the disc B is at the room temperature. The water bath E acts as a hot junction. It is heated and at various temperatures of the hot junction, the corresponding deflections in the galvanometer are noted. A graph between the difference of temperature of the hot junction and the room temperature along the Y-axis and galvanometer deflection along X-axis is plotted (Fig. 3.3). From the graph,

$$\frac{dT}{d\theta} = \tan \alpha = \frac{AB}{BC} \quad ...(i)$$

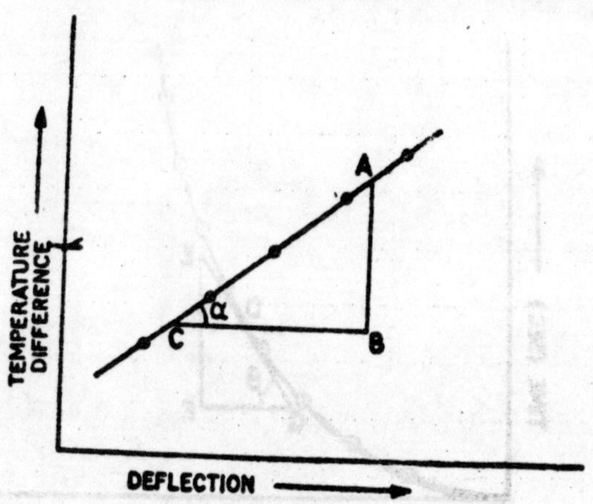

Fig. 3.3

2. The disc is completely covered with C and steam is passed into the chamber. After some time, the thermometers T T show constant temperature. The bath E is kept at room temperature. With the help of the handle B, the cover C is tilted so that the upper surface of the disc B reccives the radiations from the enclosure. The deflections in the galvanometer are observed after equal intervals of time (say 10 seconds). A graph is plotted between time and deflection (Fig. 3.3). A tangent is drawn on the curve at a point D.

$$\frac{dt}{d\theta} = \tan \beta = \frac{EF}{GF} \quad ...(ii)$$

Theory

Let, at any instant, the temperature of the enclosure and the disc be T_1 and T_2 (degrees Kelvin) respectively. The disc will absorb more heat from the surroundings and radiate less heat to the surroundings. Its temperature will rise. From Stefan's law,

$$R_1 = \sigma T_1^4 \text{ and } R_2 = \sigma T_2^4$$

$$\therefore \quad (R_1 - R_2) = \sigma (T_1^4 - T_2^4) \quad ...(iii)$$

Here R_1 is the amount of heat radiation absorbed per unit area per second by the disc and R_2, is the amount of heat radiation emitted per unit area per second by the disc.

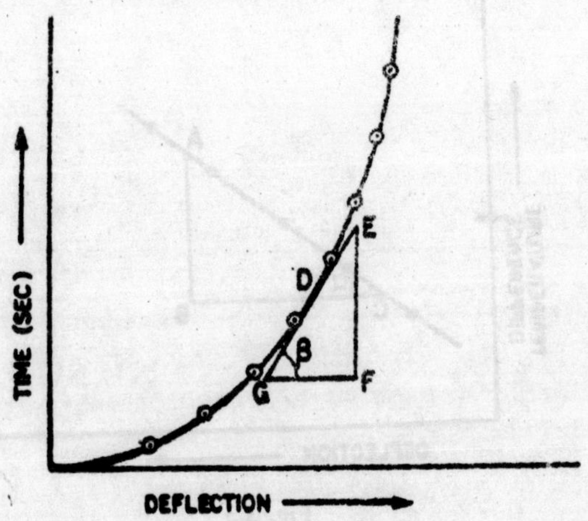

Fig. 3.4

Let the mass of the disc be m, specific heat S, rate of rise of temperature $\frac{dT}{dt}$, and area of the upper surface A.

Then $$\frac{(R_1 - R_2)A}{J} = mS\frac{dT}{dt}$$

$$\frac{\sigma(T_1^4 - T_2^4)A}{J} = mS\frac{dT}{dt}$$

$$\sigma = \frac{JmS}{(T_1^4 - T_2^4)} \times \frac{dT}{dt} \qquad \text{...(iv)}$$

To find $\frac{dT}{dt}$, equations (i) and (ii) are used.

$$\frac{dT}{dt} = \frac{dT}{d\theta} = \frac{d\theta}{dt} = \frac{\tan\alpha}{\tan\beta}$$

To find T_2, the deflection in the galvanometer corresponding to the point D on graph in Fig. 3.4 is noted and for this deflection the temperature difference from the graph (Fig. 3.3) is noted. To this reading add the room temperature and find T, in degrees Kelvin.

Substituting these values in equation (iv)

$$\sigma = \frac{JmS \tan \alpha}{A(T_1^4 - T_2^4) \tan \beta}$$

Hence σ can be calculated.

3.6 DISAPPEARING FILAMENT OPTICAL PYROMETER

The instrument was designed by Holborn, Kurlbaum and Henning.

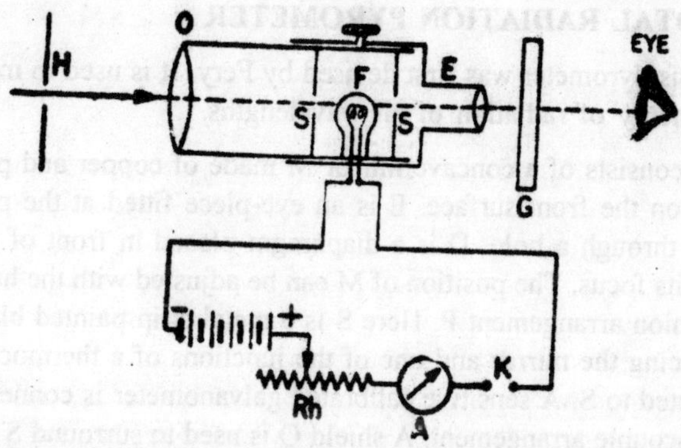

Fig. 3.5

It consists of a telescope and its cross wires are replaced by the filament F of the electric lamp (Fig. 3.5). The filament is heated with the help of a battery and the current passing through it can be adjusted with the help of a rheostat. S_1, S are two diaphragms and O and E are the objective and the eye-piece respectively. The filament is viewed through the eye-piece by placing red glass filter G between the eye and the eye-piece.

The radiations passing through the hole H are allowed to enter the objective of the telescope. By adjusting the position of the telescope, the hole is focussed on the filament F. The filament is visible against a background. The current through the filament is adjusted such that it begins to glow and the filament *becomes invisible* against the image of a bright hole. The red glass filter G helps in obtaining perfect matching for monochromatic radiation. The current passing through the filament, when it *disappears*, is noted. Since the brightness or heat energy emitted per sq cm is the same in the two cases, the temperature of the source from which radiations are allowed to enter the telescope is the same as that of the filament.

Initially, the instrument is calibrated by talking sources at known temperatures and the corresponding currents passing through the filament are noted. A graph is plotted between the current and the temperature. From the graph, for a particular value of current passing through the filament (for disappearing position) the corresponding unknown temperature can be determined. The ammeter can also be calibrated to measure the temperatures directly.

3.7 TOTAL RADIATION PYROMETER

This pyrometer was first devised by Fery. It is used to measure the total energy of radiation of all wavelengths.

It consists of a concave mirror M made of copper and plated with nickel on the front surface. E is an eye-piece fitted at the pole of the mirror through a hole. D is a diaphragm placed in front of the mirror and at its focus. The position of M can be adjusted with the help of rack and pinion arrangement P. Here S is a metal strip painted black on the side facing the mirror and one of the junctions of a thermocouple T is connected to S. A sensitive calibrated galvanometer is connected in the thermocouple arrangement. A shield Q is used to surround S so that the radiation is not incident directly on S (Fig. 3.6).

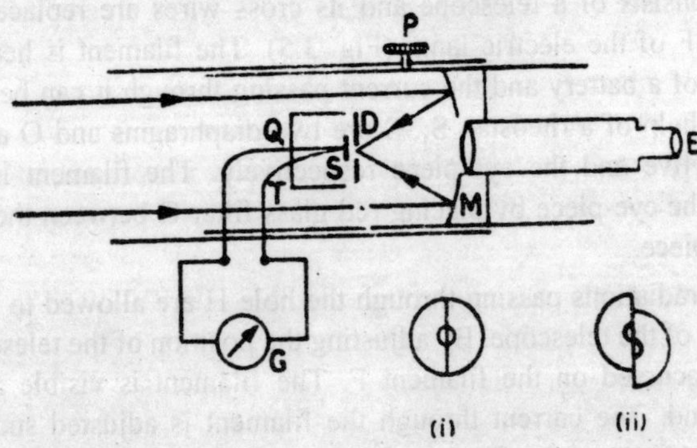

Fig. 3.6

When the radiation is incident on the mirror, the focussed image falls on the opening of D. Due to the incident radiations falling on S, the galvanometer shows a deflection. In order to focus the image accurately, the diaphragm D is made of two semicircular mirrors inclined at a small angle to each other. When the diaphragm D is viewed through the eyepiece, it appears to be circular as shown in Fig. 3.6 (i), if the focussing is accurate. If focussing is not proper, the two halves appear displaced as shown in Fig. 3.6 (ii).

The image formed on D must completely cover the aperture. The reading of the galvanometer of the thermocouple is independent of the distance of the source provided the aperture of D is completely filled by the image of the source (furnace etc.). The reading of the galvanometer depends only on the intensity of the image. If the distance of the source is doubled, the amount of radiation received by the mirror becomes one-fourth. But the area of the image is also reduced to one-fourth. Thus the intensity of the image remains the same.

Calibration

If the temperature of the source is T and that of the receiving instruments is T_0, the deflections of the galvonometer are not proportional to $(T^4 - T_0^4)$. This is because T is large as compared to T_0. The instrument is directly calibrated with source at various temperatures. (The temperature of the source is measured with a standard thermocouple or an optical pyrometer). A graph is obtained between the logarithm of the deflection of the galvanometer and the logarithm of the absolute temperature T,

$$\log \theta = K \log T.$$

The graph will be a straight line and this method gives better results. Accurate experiments show that due to the conduction along the wires of the thermocouples and stray e.m.f. generated, Stefan's fourth power law is not strictly applicable. The power of T varies from 3.8 to 4.2.

Rotating Sector Device

To measure the temperature over a wide range, a rotating sector device is used (Fig. 3.7). It consists of a circular disc with a sector of suitable angle θ cut from the disc. The disc is rotated about the axis of the pyrometer and it limits the amount of radiation incident on the mirror.

Let T be the temperature of the source and T_1 the temperature measured with the instrument. Then

$$\frac{T^4}{T_1^4} = \frac{360}{\theta}$$

$$T^4 = \frac{T_1^4(360)}{\theta}$$

$$T = T_1 = \left[\frac{360}{\theta}\right]^{1/4}$$

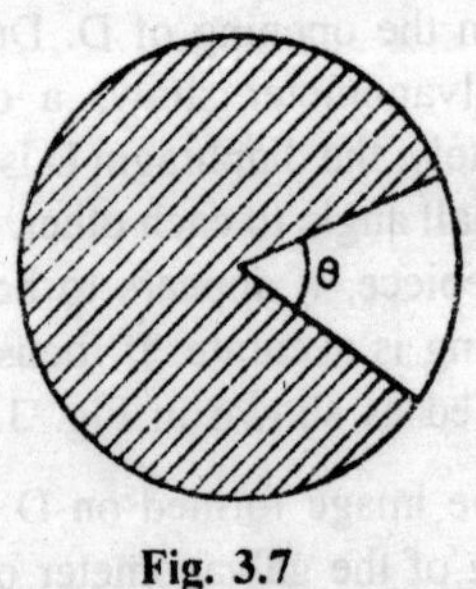

Fig. 3.7

3.8 DISTRIBUTION OF ENERGY IN THE SPECTRUM OF A BLACK BODY

Lummer and Pringsheim investigated the distribution of energy amongst the different wavelengths of a thermal spectrum of a black body radiation (full emitter).

The experimental arrangement consists of a carbon tube heated electrically (Fig. 3.8). The radiations from this tube are allowed to be incident on a reflector M_1 through a slit S_1. The parallel reflected beam is dispersed by a fluorspar prism. The dispersed beam is

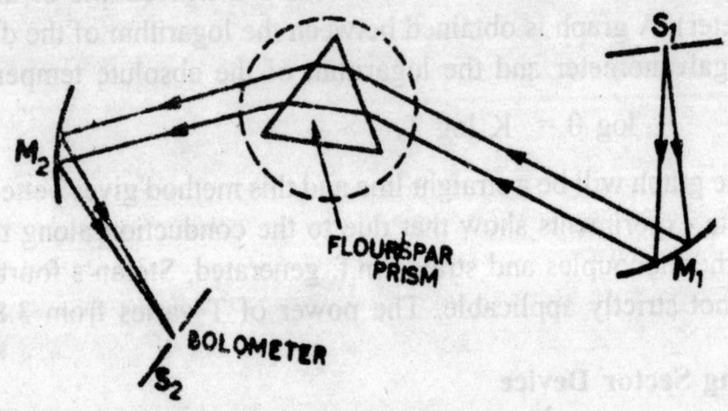

Fig. 3.8

focussed on a line bolometer with the help of the mirror M_2. The intensity of the radiations corresponding to different wavelengths

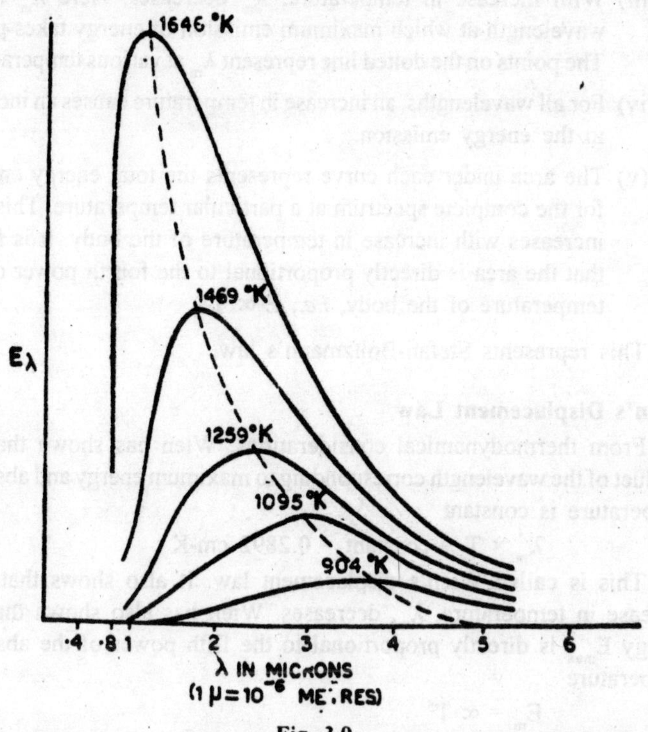

Fig. 3.9

is measured with the bolometer. The deflection in the galvanometer (of the bolometer) is proportional to the energy of the heat radiations.

To find the energy distribution for different wavelengths the fluorspar prism is gradually rotated.

The results obtained by Lummer and Pringsheim are shown in Fig. 3.9. Each curve represents the intensity of the radiations (E_λ) with wavelength for a given temperature of the source.

Results:

(i) At a given temperature, the energy is not uniformly distributed in the radiation spectrum of a hot body.

(ii) At a given temperature the intensity of heat radiations increases with increase in wavelength and at a particular wavelength its value is maximum. With further increase in wavelength the intensity of heat radiations decreases.

(iii) With increase in temperature, λ_m decreases. Here λ_m is the wavelength at which maximum emission of energy takes place. The points on the dotted line represent λ_m at various temperatures.

(iv) For all wavelengths, an increase in temperature causes an increase in the energy emission.

(v) The area under each curve represents the total energy emitted for the complete spectrum at a particular temperature. This area increases with increase in temperature of the body. It is found that the area is directly proportional to the fourth power of the temperature of the body, *i.e.*, $E \propto T^4$.

This represents Stefan-Boltzmann's law.

Wien's Displacement Law

From thermodynamical considerations, Wien has shown that the product of the wavelength corresponding to maximum energy and absolute temperature is constant

$$\lambda_m \times T = \text{constant} = 0.2892 \text{ cm-K} \quad \text{...(i)}$$

This is called Wien's displacement law. It also shows that with increase in temperature, λ_m, decreases. Wien has also shown that the energy E_{max} is directly proportional to the fifth power of the absolute temperature

$$E_m = \propto T^5$$

$$E_m = \text{constant} \times T^5 \quad \text{...(ii)}$$

Wien deduced the radiation law for the energy emitted at a given wavelength A and at a temperature T.

$$E_\lambda = \lambda^{-5} f(\lambda T) \quad \text{...(iii)}$$

With certain assumptions Wien was able to deduce the relation

$$f(\lambda T) = C_1 e^{-\frac{C_2}{\lambda T}}$$

$$\therefore \quad E_\lambda = C_1 \lambda^{-5} e^{-\frac{C_2}{\lambda T}} \quad \text{...(iv)}$$

Here C_1 and C_2 are constants.

Equation (iv) represents the Wien's law of distribution of energy.

Rayleigh-Jean's Law

The energy distribution in the thermal spectrum according to Rayleigh, is given by the formula

$$E_\lambda = \frac{8\pi kT}{\lambda^4} \quad ...(v)$$

Here k is the Boltzmann's constant.

The experimental results of Lummer and Pringsheim in the infrared region, however, show that the Wien's law holds good only in the region of shorter wavelengths. It does not hold good at longer wavelengths. The Rayleigh-Jean's law holds good in the region of longer wavelengths but not for shorter wavelengths. This was shown by Rubens and Kurlbaum. Thus Wien's law and Rayleigh-Jean's law do not precisely agree with the experimental results.

Planck's Law

Planck (1901) was able to derive a theoretical expression for the energy distribution on the basis of quantum theory of heat radiations. According to quantum theory, energy is emitted in the form of packets or quanta called photons. Each photon has an energy hν where h is the Planck's constant and ν is the frequency of radiation. According to this theory, the body does not emit energy continuously but only in-certain multiples of the fundamental frequency of the *resonator* (energy emitter). As the energy of a photon is hν, the energy emitted is equal to hν, 2hν, 3hν....... etc. He deduced the formula,

$$E_\lambda = \frac{8\pi hc}{\lambda^5\left[e^{\frac{h\nu}{kT}} - 1\right]} \quad ...(vi)$$

Here c is the velocity of electromagnetic waves (= 3×10^8 metres per second).

Equation (vi) agrees with the experimental results.

(i) For shorter wavelengths

$$e^{\frac{h\nu}{kT}} \geq 1$$

Front equation (vi),

$$\therefore \quad E_\lambda = 8\pi hc\lambda^{-5}e^{-\frac{h\nu}{kT}}$$

$$E_\lambda = C_1\lambda^{-5}e^{-\frac{C_2}{\lambda T}} \quad ...(vii)$$

where $C_1 = 8phc$ and $C_2 = \frac{hc}{\lambda}$

Equation (vii) represent's Wien's radiation law.

(ii) For longer wavelengths, $\frac{h\nu}{kT}$ is small.

From equation (vi), expanding $e^{\frac{h\nu}{kT}}$ and neglecting higher powers,

$$E_\lambda = \frac{8\pi hc}{\lambda^5\left[1+\frac{h\nu}{kT}-1\right]}$$

$$= \frac{8\pi hc.\lambda kT}{\lambda^5.hc}$$

$$E_\lambda = \frac{8\pi kT}{\lambda^4} \quad \text{...(viii)}$$

Equation (viii) represents Rayleigh-Jean's Jaw.

Thus Planck's formula for the energy distribution in a thermal spectrum is applicable for all wavelengths.

3.9 SOLAR CONSTANT

The sun is the source of heat radiations and it emits heat radiations in all directions. The earth receives only a fraction of the energy emitted by the sun. The atmosphere also absorbs a part of the heat radiations and air, clouds, dust particles etc., in the atmosphere scatter the heat and light radiations falling on them. From the quantity of heat radiations received by the earth, it is possible to estimate the temperature of the sun. Therefore, to determine the value of a constant, called solar constant, certain ideal conditions are taken into consideration.

Solar Constant

It is the amount of heat energy (radiation) absorbed per minute by one sq cm of a perfectly black body surface placed at a mean distance of the earth from the sun, in the absence of the atmosphere, the surface being held perpendicular to the sun's rays.

The instruments used to measure the solar constant are called *pyrheliometers.* The heat energy absorbed by a known area in a fixed time is found with the help of the pyrheliometers. To eliminate the effects of absorption by the atmosphere, the value of the solar constant is found at various altitudes of the sun on the same day under similar sky conditions. If S is the observed solar constant, S_0 the true solar constant and Z the altitude (angular elevation) of the sun, then

$$S = S_0 \, a^{\sec z} \quad \text{...(i)}$$

$$\log S = \log S_0 + \sec Z \log a \quad \text{...(ii)}$$

Here a is a constant.

A graph is plotted between log S along the y-axis and sec Z along the x-axis. The graph is a straight line. Produce the graph to meet the y-axis. The intercept on the y-axis gives log S_0. From the value of log S_0, the value of S_0 the solar constant can be calculated. The value obtained varies between 1.90 and 2.60 calories per sq cm per minute.

3.10 TEMPERATURE OF THE SUN

The sun consists of a central hot pardon surrounded by the photosphere. The central portion has a temperature of the order of 10^7K. The photosphere has a temperature of about 6000 K. This temperature is also called the effective temperature of the sun. Considering the sun as a perfect black body radiator, the temperature of the sun can be calculated.

Let the mean distance of the sun from the earth be R and S the solar constant. Then, the total amount' of heat energy received by the sphere of radius R in one minute = $4\pi R^2 S$.

If r is the radius of the sun, then the amount of heat energy radiated by 1 sq cm surface of the sun in one minute

$$E = \frac{4\pi R^2 S}{4\pi r^2} = \left(\frac{R}{r}\right)^2 \times S$$

Taking $R = 148.48 \times 10^7$ km

$r = 6.928 \times 10^6$ km

The mean value of

$S = 1.94$ cals per cm^2 per minute.

$$E = \left(\frac{148.48 \times 10^7}{6.928 \times 10^5}\right)^2 \times \frac{1.94}{60} \text{ cal per second} \quad \text{...(i)}$$

Also $E = \sigma T^4$

But $\sigma = 5.75 \times 10^{-5}$ ergs per cm^2 per second

$$= \frac{5.75\times10^{-5}}{4.2\times10^{7}} \text{ cal per cm}^2 \text{ per second}$$

$$E = \frac{5.75\times10^{-5}}{4.2\times10^{7}}.T^4 \qquad \text{...(ii)}$$

Equating (i) and (ii)

$$\left(\frac{5.75\times10^{-5}}{4.2\times10^{7}}\right)T^4 = \left(\frac{148.48\times10^{7}}{6.928\times10^{5}}\right)\times\frac{1.94}{60}$$

$$T = 5730 \text{ K.} \qquad \text{...(iii)}$$

This temperature gives the effective temperature of the sun acting as a black body radiator. The actual temperature of the sun is higher than this value. The temperature of the sun is usually taken as 6000 K.

Temperature of the sun can also be calculated from Wien's displacement law,

$$\lambda_{max} T = 0.2892$$

The wavelength of the radiations for which the energy is maximum in the spectrum is 4900 × 10^{-8} cm.

Substituting the value of λ_m, the value of T comes out to be 5902 K. This value is in agreement with the accepted value. Hence, the effective temperature of the sun (photosphere) is about 6000 K.

3.11 ANGSTROM'S PYRHELIOMETER

Pyrheliometer is an instrument which is used to find the amount of incident heat radiations and the solar constant. Angstrom's pyrheliometcr consists of two identical strips A and B of blackened platinum foil (Fig. 3.10).

The strip A is exposed to the sun and B is shielded by a cover C. A thermocouple having a sensitive galvanometer with A as one junction and B as the other junction is used. The strip B can be heated by an electrical arrangement and suitable current passing through B can be adjusted with the help of a rheostat.

When both the strips A and B are shielded from the sun, their junctions are at the same temperature and the galvanometer shows no deflection. The strip A is exposed to the sun and B is kept shielded. The strip A received heat radiations 'from the sun and its temperature rises and the galvanometer shows deflection. The strip B is gradually heated by passing current through it and the current is adjusted so that there is no deflection in the galvanometer.

It means the strips A and B are again at the same temperature and they are receiving heat energy at the same rate.

Let B calories of heat be incident on one sq cm surface of the strip in one minute. The area of the plate A = A sq cm and absorption coefficient = a.

The amount of heat radiations absorbed: in one minute by the plate

$$A = H\ Aa \text{ calories} \quad ...(i)$$

Heat produced in one minute in the strip B

$$= \frac{EI \times 60}{4.2} \text{cal} \quad ...(ii)$$

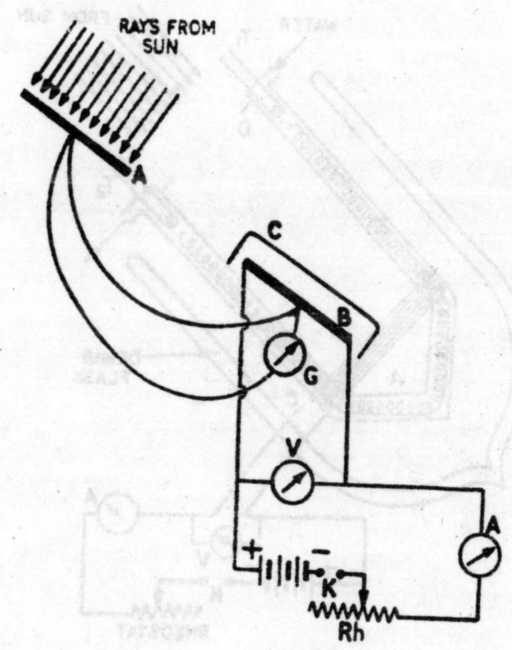

Fig. 3.10

Here S volts is the potential difference across the strip B and I amperes is the current flowing through it.

Equating (i) and (ii),

$$HAa = \frac{EI \times 60}{4.2} \quad ...(iii)$$

Hence H can be calculated.

3.12 WATER FLOW PYRHELIOMETER

This instrument was designed by Abbot and Fowle. It consists of a cylindrical doubled walled vessel A The lower side is conical in shape and is coated with platinum black on its inner side. Water can be circulated through the walls at a constant rate. T_1 and T_2, are the platinum resistance thermometers to record the temperatures of incoming and outgoing water (Fig. 3.11).

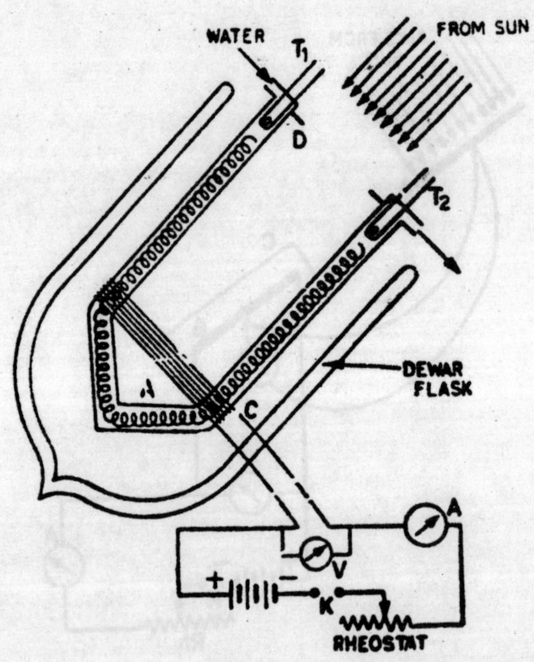

Fig. 3.11

D is a diaphragm which allows the heat radiations from the sun through a known area of cross-section. Heat radiations are allowed to enter the Tend A and the rate of flow of water is adjusted so that the thermometers T_1 and T_2, show constant temperatures (say θ_1 and θ_2). The spiral coil helps water in making good contact with the walls. After the steady stale has been reached *i.e.*, T_1 and T_2 show constant temperatures, the heat radiations from the sun are cut off.

Now a suitable current is passed through the manganin coil C, so that the same conditions are restored *i.e.*, the thermometers T_1 and T_2 show the same constant temperatures θ_1 and θ_2. If E volts is the potential difference and I amperes is the current flowing through the manganin coil, then the amount of heat produced per second $\frac{EI}{4.2}$ cal. Therefore, the amount of heat radiations received from the sun in one minute by the exposed surface of the vessel $A = \frac{EI \times 60}{4.2}$ cal. From this value, the solar constant can be calculated.

3.13 WATER STIR PYRHELIOMETER

In this type of pyrheliometer the heat radiations are received from the sun by the conical end of the vessel A, which is coated with lamp black. Vessel A is surrounded by a calorimeter D containing water. The stirrer helps in circulating water vigorously and is driven by an electric motor. The temperature of water is measured with the help of a platinum resistance thermometer P, whose wire is wound around A (Fig. 3.12). Heat radiations from the sun are allowed to enter the vessel A and the rise in temperature of water in one second is calculated. The heat radiations are cut off and a suitable current is passed through the manganin coil C for a known time so that the rise in temperature of water in one second is the same.

If E volts is the potential difference and I amperes is the current passing through the manganin coil, then the amount of heat produced per second $= \frac{EI}{4.2}$ cal. Therefore, the amount of heat radiations received from the sun in one minute by the exposed surface of the vessel $A = \frac{EI \times 60}{4.2}$ cal. From this value, the solar constant can be calculated.

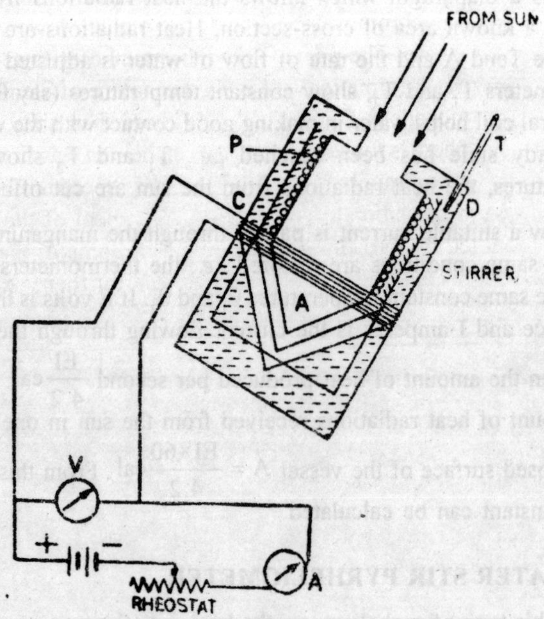

Fig. 3.12

3.14 SOLAR SPECTRUM

The radiation received from the sun is similar to that of a perfectly black body. In a solar spectrum the wavelength corresponding to maximum energy is about 5000 A for a surface temperature of 5750 K. The solar spectrum consists of a large number of dark lines called *Fraunhofer lines*. These lines were first observed by Wollaston in 1802 and later studied by Fraunhofer in 1814. These lines are observed in the complete range of the spectrum viz., ultra-violet, visible and infrared. Fraunhofer measured the wavelengths of many of these lines accurately and found that they occupied exactly the same positions as the bright lines emitted by different gases and vapours.

According to Kirchhoff's law of radiation, any substance at a lower temperature will absorb the radiations of those wavelengths which it will emit when excited by electric discharge. Thus the solar spectrum is an absorption spectrum and the Fraunhofer lines are the absorption lines of relatively cooler gases and vapours present in the earth's atmosphere and the sun's outer atmosphere. The central portion of the sun is called the photosphere which is at a very high temperature. Surrounding the photosphere is the chromosphere which is at a much lower temperature than the photosphere. The positions of the Fraunhofer lines in the solar spectrum give the information regarding the nature of elements present in the sun's atmosphere. The presence of more than sixty elements in the sun's atmosphere is found this way. In fact, helium was first discovered by the study of the Fraunhofer lines in the solar spectrum before it had been isolated in the laboratory. The most prominent of the Fraunhofer lines are denoted by the letters of the alphabet. Some of the lines, with their wavelengths and the elements responsible for their absorption are given below:

Line	Element	Wavelength in Å
A	Atmospheric oxygen	7594
B	Atmospheric oxygen	6867
C	Hydrogen	6563
D_1	Sodium	5896
D_2	Sodium	5890
F	Hydrogen	4861
G_1	Hydrogen	4341
H	Calcium	3969
K	Calcium	3934

3.15 INFRARED SPECTRUM

Extending on either side of the visible spectrum, there are invisible radiations which do not cause the sensation of sight. The radiations beyond the red end of the visible spectrum are called infrared radiations and their wavelengths extend up to 400,000 Å. Sun is a powerful source of infrared radiation. Beyond the violet end of the visible spectrum, the radiations extending up to a wavelength of 100 Å are called ultraviolet radiations.

The heating effect of the infrared radiations is used in measuring the wavelength of the radiation. An infrared ray spectrometer is shown in Fig. 3.13. Light from a strong source of light such as an electric arc is rendered parallel by reflection from a concave stainless steel mirror M_1. This parallel beam is refracted through the rock salt prism P arid the emergent dispersed beam after reflection from the mirrors M_1 and M_3 is incident on a thermopile or a bolometer. M_2 is a plane mirror and M_3 is a concave mirror. Wien's law of radiation is given by

$$\lambda_m T = 0.2892$$

where T is the absolute temperature and λ_m is the wavelength of the radiation. The thermopile readings help in the calculation of temperature and from the temperature, λ can be calculated.

As infrared radiations are not absorbed by air or thick frog, infrared ray photographs can be taken over long distances of fog and mist where visible light cannot penetrate. For this, specially designed photographic plates are used with suitable filters. A solution of iodine in alcohol is a suitable filter, because this transmits the infrared radiations and absorbs the visible light. In the World War II, infrared photography played a very useful part in detecting objects in the dark through mist, fog and clouds. Infrared radiations have a wide application in the field of medicine, industry etc. Infrared radiations can penetrate deep into the human body and by their property of heating can dilate the blood vessel at the portion exposed to the radiations. This enables increased flow of blood.

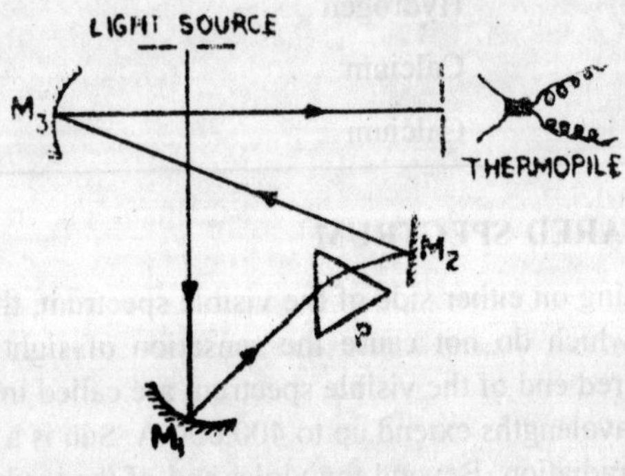

Fig. 3.13

3.16 ULTRAVIOLET SPECTRUM

The spectrum that coven the wavelengths from 4000Å to 100Å is called the ultraviolet spectrum. Sun is a natural source of light rich in ultraviolet radiations. An electric are of carbon, iron or other materials, mercury vapour lamps, discharge of electricity through hydrogen contained in quartz tubes are some of the artificial sources that give ultraviolet radiations.

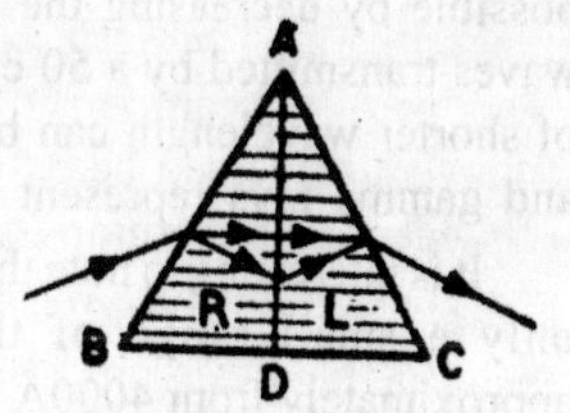

Fig. 3.14

Ordinary glass absorbs the ultraviolet radiations. Hence, quartz lenses and prisms are used. But quartz is a doubly refracting material. If a single prism of quartz is used, due to the property of double refraction of the prism, two images of the slit (ordinary and extraordinary) corresponding to a single wavelength are observed. This will reduce the sharpness of each image. To compensate for this, the collimating lens is made from right-handed quartz and the telescope objective is made from left-handed quartz. Similarly, the prism used for the dispersion of the incident beam consists of two halves. The two halves are held together by glycerine. One half is made from right-handed quartz and the other half is made from left-handed quartz.

In Fig. 3.14, ABD is a right-angled prism of right-handed quartz and ADC is of left-handed quartz. For recording an ultra- violet spectrum for wavelengths shorter than 1.200Å, a concave reflection grating is used. The diffracted beam is photographed. The grating and the photographic plate are enclosed in a metal chamber which is evacuated.

Ultra-violet radiations have a variety of applications. Sterilisation of rooms in which blood plasma, drugs, vaccines etc. are prepared and sealed in the containers is done by ultraviolet radiations. Drugs, poisons, dyes etc. fluoresce under the action of ultraviolet rays. The resolving power of a microscope is increased when ultraviolet light is used for illumination. Fluorescent tubes depend on the principle of fluorescence effected by ultraviolet radiations.

3.17 ELECTROMAGNETIC SPECTRUM

Visible spectrum includes those wavelengths which can stimulate the sense of sight. But there is no basic difference between light waves and the electromagnetic waves produced by electrical oscillating circuits. The term electromagnetic spectrum is used for the range of wavelengths from 10^4 metres to 1Å (10^{-8} cm). There is in fact no limit to the production of electromagnetic waves of very long wavelengths. The frequency of an alternating current generator can be made as low as possible by decreasing the speed of the generator. The wavelength of waves transmitted by a 50 cycle transmission line is 5×10^8 cm. Waves of shorter wavelength can be produced by electrical oscillators. X-rays and gamma rays represent the waves of very short lengths.

It is interesting to note that the visible range of the spectrum comprises only a small range of the electromagnetic spectrum extending approximately from 4000Å in the extreme violet region to 8000Å in the extreme red. Beyond the violet region of the visible spectrum is the ultraviolet, the X-rays and γ-rays. Beyond the red end of visible spectrum are the short radio waves and radio broadcast waves.

3.18 BOY'S RADIOMICROMETER

It is a combination of a moving coil galvanometer and a thermocouple. It consists of a single loop of silver or copper wire 4, to the lower end of which two strips of antimony and bismuth are attached. The lower ends of these strips are soldered to a copper disc coated with lamp black. When the heat radiations fall on the disc, thermoelectric current is produced in antimony-bismuth thermocouple and the current flows through the silver wire as shown in Fig. 3.15. The deflection produced in the galvanometer can be measured with a lamp and scale arrangement. This instrument is very sensitive and can detect even feeble heat radiations.

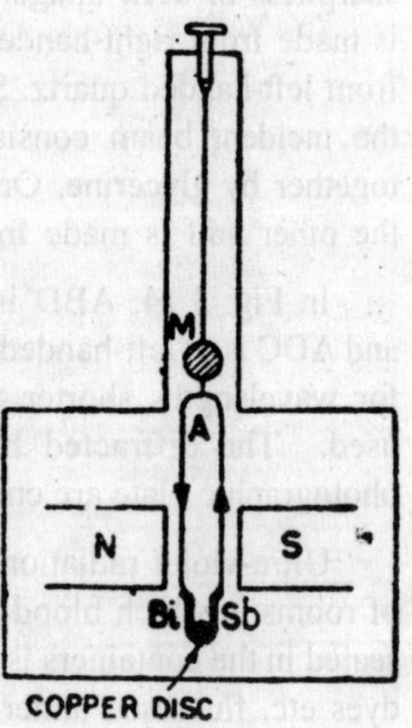

Fig. 3.15

3.19 CONDUCTION

Conduction is the process in which heat is transmitted from one point to the other through the substance without the actual motion of the particles. When one end of a metal bar is heated, the molecules at the hot end vibrate with higher amplitude (kinetic energy) and transmit the heat energy from one particle to the next and so on. However, the particles remain in their mean positions of equilibrium. This process of conduction is prominent in the case of solids. The property of transmission of heat has been used in Davy's safety lamp. Materials having less conductivity *e.g.*, granite, brick walls etc. are used in the construction of a cold storage, furnace of a boiler etc. The space between the two walls of a thermos flask is evacuated because vacuum is a poor conductor of heat. The air enclosed in the woollen fabric helps in protecting us from cold, because air is a poor conductor of heat.

3.20 CONVECTION

Convection is the process in which heat is transmitted from one place to the other by the actual movement of the heated particles. It is prominent in the case or liquids and gases. Land and sea breezes and trade winds are formed due to convection.

Convection plays an important part in ventilation, gas filled electric lamps and heating of buildings by hot water circulation.

3.21 RADIATION

Radiation is the process in which heat a transmitted from one place to the other directly without the necessity of the intervening medium. We get heat radiations directly, from the sun without affecting the intervening medium. Heat radiations can pass through vacuum. Their properties are similar to light radiations. Heat radiations also form a part of the electromagnetic spectrum.

3.22 COEFFICIENT OF THERMAL CONDUCTIVITY

Consider a cube of side x cm and area of each face A sq cm. The opposite faces of the cube are maintained at temperatures θ_1 and θ_2 where $\theta_1 > \theta_2$. Heat is conducted in the direction of the fall of temperature. Quantity of heat conducted across the two opposite faces

$$Q \propto A$$

$$Q \propto (\theta_1 - \theta_2)$$

$$Q \propto t$$

$$\propto \frac{1}{x}$$

$$Q \propto \frac{A(\theta_1 - \theta_2)}{x}$$

$$Q = \frac{KA(\theta_1 - \theta_2)t}{x}$$

Here K is a constant called the coefficient of thermal conductivity of the material of the cube.

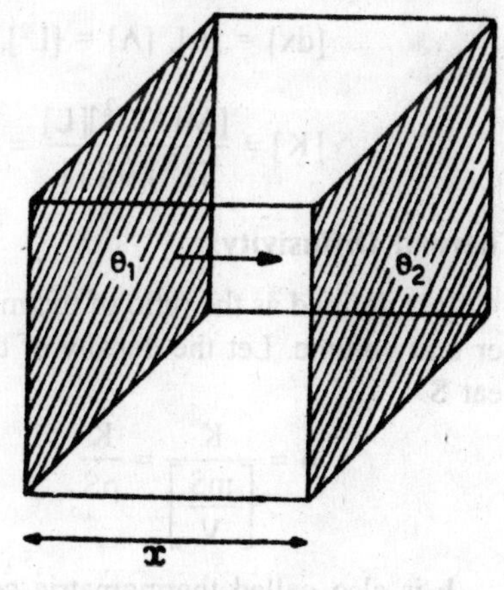

Fig. 3.16

If $A = 1$ sq cm, $(\theta_1 - \theta_2) = 1°C$,

$t = 1$ s, $x = 1$ cm

then $Q = K$.

Therefore, the coefficient of thermal conductivity is defined as the amount of heat flowing in one second across the opposite faces of a cube of side one cm maintained at a difference of temperature of 1°C.

Temperature Gradient

The quantity $\frac{(\theta_1 - \theta_2)}{x}$ represents the rate of fall of temperature with respect to distance. The quantity $d\theta/dx$ represents the rate of change of temperature with respect to the distance. As temperature decreases with increase in distance from the hot end, the quantity $d\theta/dx$ is negative and is called the temperature gradient.

$$\therefore \quad Q = KA\left[\frac{d\theta}{dx}\right]t$$

Dimensions of K

$$KA = -\frac{Q}{A\frac{d\theta}{dx}t} = -\frac{Qdx}{Ad\theta t}$$

Q represents energy and its dimensions are,

$$[Q] = [MLT^{-2}],$$

$$[dx] = [L],\ [A] = [L^2],\ [d\theta] = [\theta],\ [t] = [T]$$

$$[K] = \frac{[ML^2T^{-2}][L]}{[L^2][\theta][T]} = [MLT^{-3}\theta^{-1}].$$

Thermal Diffusivity

It is defined as the ratio of thermal conductivity to thermal capacity per unit volume. Let the density of the substance be ρ and its specific heat S.

$$h = \frac{K}{\left[\frac{mS}{V}\right]} = \frac{K}{\rho S}$$

h is also called thermometric conductivity.

3.23 RECTILINEAR FLOW OF HEAT ALONG A BAR

Consider a bar of uniform area of cross-section which is heated at one end. Heat is flowing along the length of the bar. Consider two planes P_1 and P_2 perpendicular to the length of the bar at distances x and x + δx; from the hot end. The temperature gradient at the plane $P_1 = \frac{d\theta}{dx}$. Here θ is the excess of temperature (above the surroundings) of the bar at P_1. The excess of temperature at

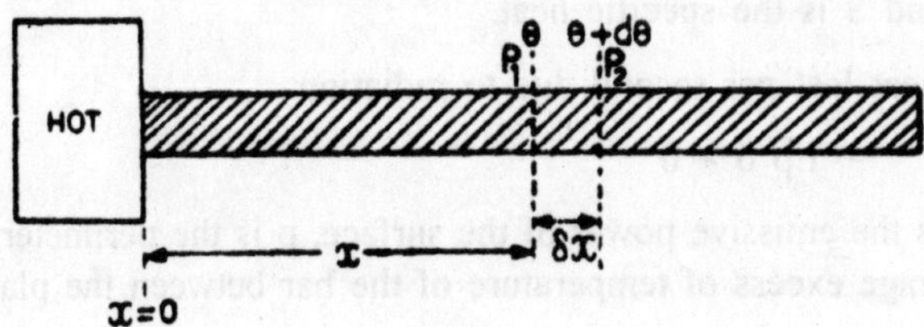

Fig. 3.17

$$P_2 = \theta + \frac{d\theta}{dx}\delta x$$

The temperature gradient at $P_2 = \frac{d}{dx}\left(\theta + \frac{d\theta}{dx}\delta x\right)$

Heat flowing through P_2 in one second

$$Q_1 = -KA\frac{d\theta}{dx} \qquad \text{...(i)}$$

Heat flowing through P_2 in one second

$$Q_2 = -KA\frac{d}{dx}\left(\theta + \frac{d\theta}{dx}\delta x\right) \qquad \text{...(ii)}$$

Heat gained per second by the rod between the planes P_1 and P_2

$$Q = Q_1 - Q_2$$

$$Q_1 = -KA\frac{d\theta}{dx} + KA\frac{d}{dx}\left(\theta + \frac{d\theta}{dx}\delta x\right)$$

$$Q = -KA\frac{d^2\theta}{dx^2}\delta x \qquad \text{...(iii)}$$

Before the Steady State is Reached

The quantity of heat Q is used in two ways before the steady state is reached. Partly the heat is used to raise the temperature of the rod and the rest is lost due to radiation. Let the rate of rise of temperature of the bar be $\frac{d\theta}{dt}$. The heat used per second to raise the temperature of the rod

$$= \text{mass} \times \text{specific heat} \times \frac{d\theta}{dt}$$

$$= (A \times \delta x)\rho \times S\frac{d\theta}{dt} \quad \text{...(iv)}$$

where A is the area of cross-section of the rod, ρ is the density of the material and S is the specific heat.

The heat lost per second due to radiation

$$= Ep\ \delta \times \theta \quad \text{...(v)}$$

where E is the emissive power of the surface, p is the perimeter and θ is the average excess of temperature of the bar between the planes P_1 and P_2.

$$Q = A\delta x\rho \times S\frac{d\theta}{dt} + Ep.\delta x.\theta$$

Substituting the value of Q from equation (iii)

$$KA\frac{d^2\theta}{dx^2}\delta x = A\delta x\rho S\frac{d\theta}{dt} + Ep\delta x\theta$$

$$\frac{d^2\theta}{dx^2} = \frac{\rho S}{K}\frac{d\theta}{dt} + \frac{Ep}{KA}\theta \quad \text{...(vi)}$$

This is the general equation that represents the rectilinear flow of heat along a bar of uniform area of cross-section.

Special Cases

1. *When heat lost by radiation is negligible*: When the rod is completely covered by insulating materials, the heat lost by radiation Ep.δx.θ is zero. In that case the total heat gained by the rod is used to raise the temperature of the rod. From eqn. (vi)

$$\frac{d^2\theta}{dx^2} = \frac{\rho S d\theta}{K dt} = \frac{1 d\theta}{h dt} \quad \text{...(vii)}$$

Here, $\frac{K}{\rho S}$ = h, the thermal diffusivity of the rod.

2. After the steady state is reached. When the steady state is reached, the rod does not require any further heat to raise its temperature. At this stage,

$$\frac{d\theta}{dt} = 0$$

From equation (vi)

$$\frac{d^2\theta}{dx^2} = \frac{Ep}{KA}\theta$$

Taking $$\frac{Ep}{KA} = \mu^2$$

$$\frac{d^2\theta}{dx^2} = \mu^2\theta \qquad \text{...(viii)}$$

The general solution of this equation is

$$\theta = Ae^{+\mu x} + Be^{-\mu x} \qquad \text{...(ix)}$$

where A and B axe two unknown constants to be determined from the boundary conditions of the problem.

If the bar is sufficiently long, we can assume that under steady state, no heat is lost from the free end of the bar. This is due to the fact that nearly the whole of the heat energy is lost from the sides as radiation and the free end will be at the temperature of the surroundings.

(a) *Suppose the bar is of infinite length* : Let the excess of temperature above the surroundings of the hot end be θ_0 and of the other end be zero.

At $$x = 0, \theta = \theta_0$$

From equation (ix)

$$x = A + B$$

At $$x = \infty, \theta = 0$$

From equation (ix),

$$0 = Ae^{\infty}$$

But e^{∞} cannot be zero, therefore

$$A = 0$$

Taking $$A = 0$$

$$\theta_0 = B$$

Substituting the values of A and B in equation (ix)

$$\theta = \theta_0 e^{-\mu x} \qquad ...(x)$$

This equation represents the excess of temperature of a point at a distance x from the hot end after the steady state is reached and this represents an exponential curve. This equation is useful in Ingen-Hausz experiment.

(b) *Suppose the bar is sufficiently long and is of finite length L.* Then the boundary conditions are

At $x = 0$

$\theta_0 = \theta$

and $\frac{d\theta}{dx} = 0$ at $x = L$

The values of A and B in this case are

$$A = \frac{\theta}{1 + e^{+2\mu L}}$$

and $$A = \frac{\theta}{1 + e^{+2\mu L}}$$

The solution of equation (ix) will be

$$\theta = \left[\frac{e^{\mu x}}{1 + e^{2\mu L}} + \frac{e^{-\mu x}}{1 + e^{-2\mu L}}\right]$$

3.24 INGEN-HAUSZ EXPERIMENT

Ingen-Hausz devised an experiment to compare the thermal conductivities of several materials.

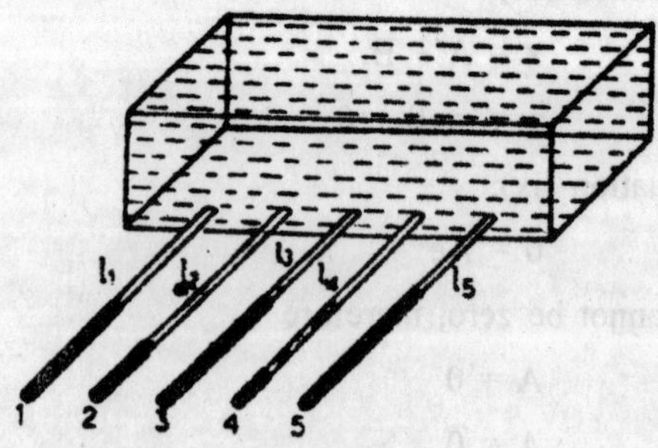

Fig. 3.18

Take a box containing hot water or oil. Rods of different materials of the same length and area of cross-section (same size and shape) are taken. Their outer surfaces are electroplated with the same material and are equally polished This is done such that the emissive power E for all the rods is the same. Rods are fixed at the base of the vessel (Fig. 3.18) and their portions outside the vessel are coated uniformly with wax. Hot water or oil is put into the vessel and the wax starts melting. It is found that the wax melts up to different lengths on different rods.

Let the lengths up to which the wax has melted be l_1, l_2, l_3.... etc., θ_0 be the excess of temperature of the hot bath above the room temperature and θ_m the excess of temperature of the melting point of wax above the room temperature.

As,
$$\theta = \theta_0 e^{-\mu x}$$

For the first rod,
$$\theta_m = \theta_0 e^{-\mu 1 l 1}$$

For the second rod,
$$\theta_m = \theta_0 e^{-\mu 2 l 2}$$

For the third rod,
$$\theta_m = \theta_0 e^{-\mu 3 l 3}$$

....

$$\therefore \quad \theta_m = \theta_0 e^{-\mu 1 l 1} = \theta_0 e^{-\mu 2 l 2}$$
$$= \theta_0 e^{-\mu 3 l 3}$$

$$\therefore \quad \mu_1 l_1 = \mu_2 l_2 = \mu_3 l_3 = \ldots$$

But
$$\mu_1 = \sqrt{\frac{Ep}{K_1 A}}$$

$$\mu_2 = \sqrt{\frac{Ep}{K_2 A}}$$

$$\mu_3 = \sqrt{\frac{Ep}{K_3 A}}$$

Here, E, p and A are the same for all the rods

$$\therefore \quad l_1 = \sqrt{\frac{Ep}{K_3 A}} = l_2 \sqrt{\frac{Ep}{K_2 A}}$$

$$= l_3 \sqrt{\frac{Ep}{K_3 A}} = \ldots$$

$$\therefore \quad \frac{l_1}{\sqrt{K_1}} = \frac{l_2}{\sqrt{K_2}}$$

$$= \frac{l_3}{\sqrt{K_3}} \text{ constant}$$

or
$$\frac{K_1}{l_1^2} = \frac{K_2}{l_2^2} = \frac{K_3}{l_3^2} \text{ constant} \qquad \ldots(i)$$

Thus, the thermal conductivity of the material of the rod is directly proportional to the square of the length up to which the wax melts on the rod.

From equation (i), the thermal conductivities of any two materials can be compared. If the conductivity of one of the rods is known, the thermal conductivities of other rods can be determined. This experiment is not used for finding the thermal conductivity because the results obtained are not accurate. It is, however, useful for demonstration purposes.

3.25 SEARLE'S METHOD

XY is a rod whose coefficient of thermal conductivity is to be determined. The end X is enclosed in a steam jacket and team is

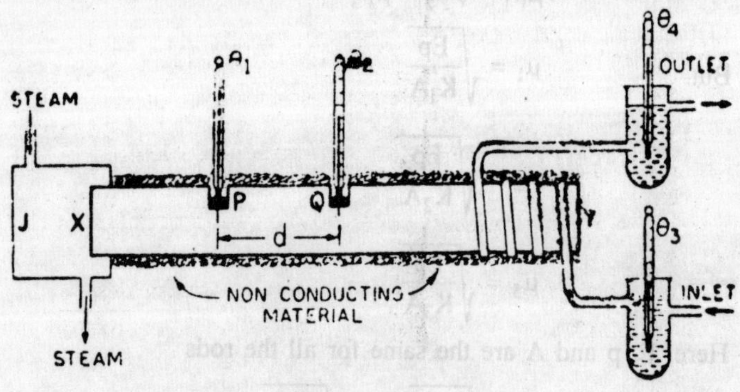

Fig. 3.19

continuously passed through it. Heat is conducted along the rod and when the steady state is reached, the thermometers θ_1, θ_2 θ_3 and θ_4 record constant temperatures. The mass of water (m grams) collected in a known interval of time (t seconds) is noted (Fig. 3.19).

When the steady state is reached, the quantity of heat passing from the section P to Q is constant and the same quantity of heat flow through the section at Y. The rod is well covered with a non-conducting material like felt, cotton etc. to avoid conduction and convection of heat to the atmosphere.

Quantity of heat that flows from the section at P to Q in one second,

$$= \frac{KA(\theta_1 - \theta_2) \times 1}{d}$$

Amount of heat gained by water in one second

$$= \frac{m(\theta_4 - \theta_3)}{t}$$

Equating (i) and (ii)

$$\frac{KA(\theta_1 - \theta_2)}{d} = \frac{m(\theta_4 - \theta_3)}{t}$$

$$\therefore \quad K = \frac{md(\theta_4 - \theta_3)}{At(\theta_1 - \theta_2)}$$

Knowing all the quantities, K can be calculated.

3.26 LEE'S METHOD FOR METALS

The thermal conductivity of a metal is different at different temperatures. This method is useful in measuring the thermal conductivity of a metal at various temperatures, preferably at low temperatures.

The apparatus consists of a copper frame C surrounded by a heating coil H_1, The specimen rod AB about 10 cm long and 0.5cm in diameter is fitted inside the frame C (Fig. 3.20). The end A of the rod is heated by a heating coil H and T_1 and T_2 are two platinum, resistance thermometers. These thermometers measure temperatures at the two points distant d apart on the rod AB. The end B is fixed in a socket inside the frame C. The whole arrangement is surrounded by a Dewar flask F.

The whole apparatus is immersed in liquid air so that its temperature is below the temperature at which the thermal conductivity is to be measured. The liquid air is removed and with the help of the heating coil H_1 the whole apparatus is brought to the desired temperature. The heating coil H is heated by passing a knowing current through it and the temperatures of T_1 and T_2, are noted, after they show constant temperatures, θ_1 and θ_2 *i.e.*, after the steady state has been reached. Let the current flowing be I amperes and potential difference across the heating coil H be E volts.

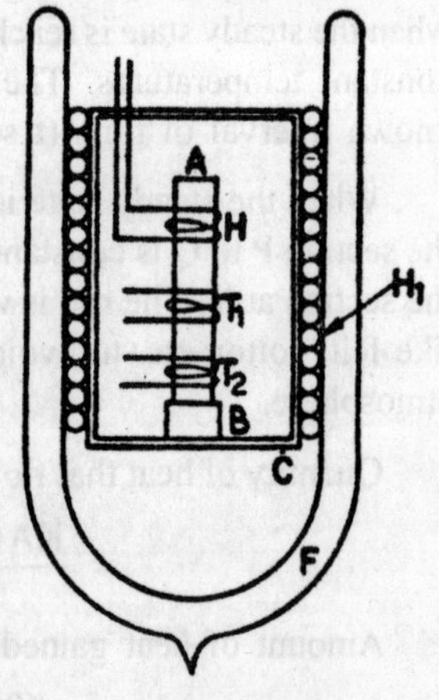

Fig. 3.20

Heat produced per second $= \dfrac{EI}{4.2}$ calories

$$\therefore \quad \frac{EI}{4.2} = \frac{KA(\theta_1 - \theta_2)}{d}$$

$$K = \frac{EId}{4.2\, A(\theta_1 - \theta_2)}$$

Hence, the thermal conductivity of the given metal at a desired temperature can be calculated.

Correction

When the end A of the rod AB is heated by the heating coil H, the copper frame conducts some heat and its temperature rises. This heat does not escape and there is backward flow of heat from the end B. Due to this reason, $(\theta_1 - \theta_2)$ will be less than the correct value. Therefore, correction is a necessity. The resistances of H_1 and H are made equal. Before performing the main experiment, the same current I is passed through H_1, without passing the current through B. The difference of temperature $d\theta_1$ between T_2 and T_1 is noted after the steady state is reached. The observation is repeated after passing the same current I through H. Let the difference of temperature be $d\theta_2$. The correction to be applied

$$= \frac{d\theta_1 - d\theta_2}{2}$$

Hence the correct difference in temperature is

$$= \frac{(\theta_1 - \theta_2) + (d\theta_1 - d\theta_2)}{2}$$

$$K = \frac{EId}{4.2\ A[(\theta_1 - \theta_2) + (d\theta_1 - d\theta_2)/2]}$$

Lee found K for various metals between −170°C and 18°C. For silver, copper, zinc and aluminium K increases with rise in temperature between −170°C and 100°C. For iron and lead, K decreases with rise in temperature. This method is very useful in finding the thermal conductivity of metals at different temperatures.

3.27 FORBES' METHOD TO FIND K

Forbes' method is used to determine the *absolute* conductivity of different metals.

Consider a long time of uniform area of cross-section heated at one end (Fig. 3.21). After the steady state is reached, the amount of heat passing per second across the cross-section at the point $B = KA\left(\frac{d\theta}{dx}\right)_B$.

Here $\left[\frac{d\theta}{dx}\right]_B$ is the temperature gradient at B. The amount of heat lost by radiation from the point B up to the end C

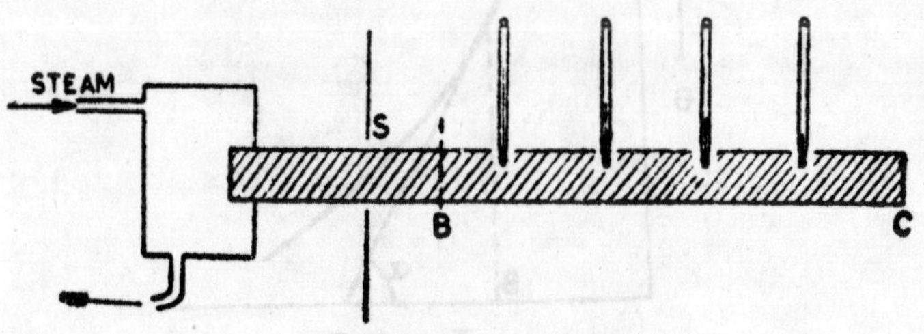

Fig. 3.21

$$= \int_B^C (\rho A dx) S \frac{d\theta}{dt}$$

$$\therefore \quad KA\left[\frac{d\theta}{dx}\right]_B = \int_B^C (\rho A dx) S \cdot \frac{d\theta}{dt}$$

$$K = \frac{\rho S \int_B^C \frac{d\theta}{dt} dx}{\left[\frac{d\theta}{dx}\right]_B} \quad ...(i)$$

To find $\int_B^C \frac{d\theta}{dt} dx$ and $\left[\frac{d\theta}{dx}\right]_B$ the experiment is divided into two parts.

(i) Static Experiment

A long rod about 2 metres in length is taken and its one end is heated (Fig. 3.21). Forbes heated the rod for about six hours. After the steady state is reached, the thermometers fixed in the rod at various positions are read. A graph is plotted between the temperature (θ) and the distance (x) from the hot end The graph is an exponential curve as shown in Fig. 3.22. Take the point B and draw a tangent to the curve corresponding to the point B.

Here, $\left[\frac{d\theta}{dx}\right]_B = \tan \alpha$

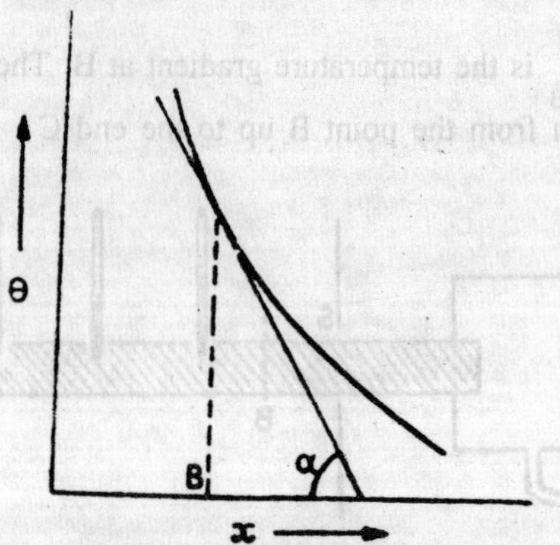

Fig. 3.22

(ii) Dynamic Experiment

The original bar or a small bar of the same material having the same area of cross-section is heated to the temperature of steam. The bar is exposed to the atmosphere and a thermometer is fixed at its middle portion (Fig. 3.23). Note the temperature of the bar after regular intervals of time (say one minute). Draw a graph between temperature and time.

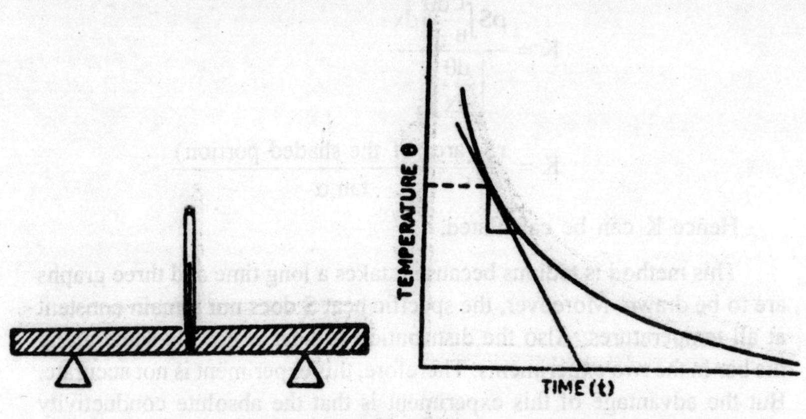

Fig. 3.23

From the graph in Fig. 3.24, for various values of θ, $\frac{d\theta}{dt}$ is determined. This is done by drawing tangents to the curve at various points on the curve.

From the two graphs given in Figs. 3.22 and 3.24, the values of $\frac{d\theta}{dt}$, corresponding to θ and x are determined. A graph is plotted between $\frac{d\theta}{dt}$ and x. The graph is produced to meet the x-axis. Corresponding to the point B, a point is located on the graph (Fig. 3.25). The area of the shaded portion $\int_B^C \frac{d\theta}{dt} dx$. The area is measured with the help of the planimeter or from the graph directly.

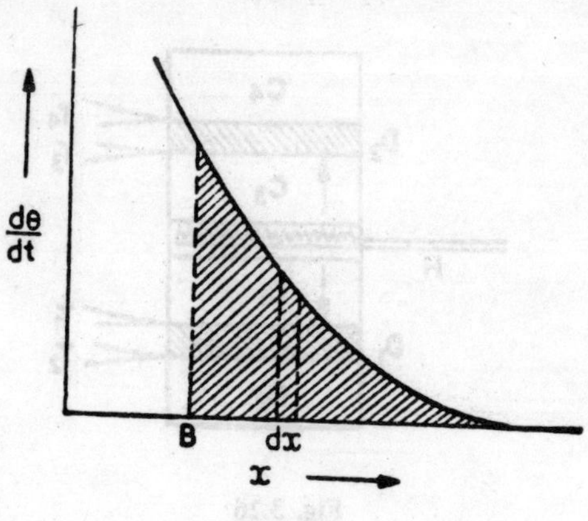

Fig. 3.25

$$K = \frac{\rho S \int_B^C \frac{d\theta}{dt} dx}{\left[\frac{d\theta}{dx}\right]_B}$$

$$K = \frac{rS \text{ (area of the shaded portion)}}{\tan \alpha}$$

Hence K can be calculated.

This method is tedious because it takes a long time and three graphs are to be drawn. Moreover, the specific heat S does not remain constant at all temperatures. Also the distribution of heat is not the same along the bar in the two experiments. Therefore, this experiment is not accurate. But the advantage of this experiment is that the absolute conductivity of the material of the rod can be determined.

3.28 LEE'S METHOD FOR BAD CONDUCTORS

This method is used to determine the thermal conductivity of poor conductors viz., rubber, glass, ebonite, wood, cork etc. The specimen is taken in the form of two thin discs D_1, and D_2 about 10 cm in diameter and 2 to 3 mm thick. The disc A is pressed between two copper plates C_1 and C_2 and D_2 is pressed between the copper plates C_3 an d C_4 (Fig. 3.26). Discs C_1 and C_3 ensure normal flow of heat through the experimental plates D_1 and D_2. H is a heater coil and T_1, T_2, T_3 and T_4 are four thermocouples used to measure temperatures. The surfaces of D_1 and D_2 are coated with glycerine so that these surfaces make good thermal contact with the copper plates.

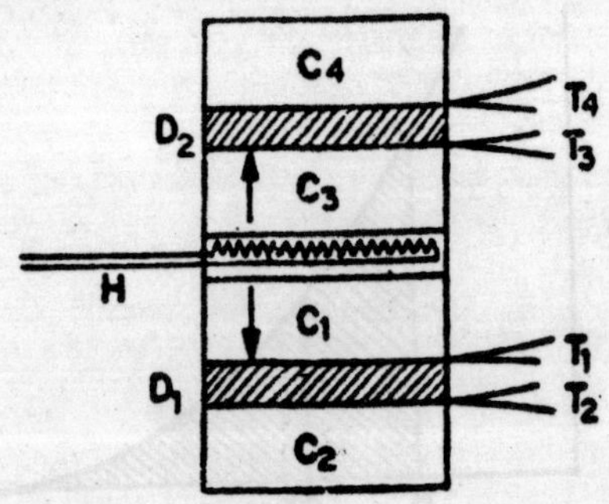

Fig. 3.26

A steady current I is passed through the heater coil H. The potential difference across the heater coil is E.

After the steady, state is reached the temperatures of the thermometers T_1, T_2, T_3 and T_4 are noted. Let the temperatures be θ_1, θ_2, θ_3 and θ_4 respectively. In the steady state, heat generated in the heater coil is lost from the surface of C_2 and C_4, and heat lost from the rims of C_1, C_2, D_1 and D_2 is negligible due to the small thickness of thc plates.

Let d_1 be the thickness of the disc D_1 and d_2 the thickness of the disc D_2.

Heat produced by heater coil in one second

$$= \frac{EI}{4.2} \text{ calories.} \qquad \text{...(i)}$$

Heat passing through D_1 and D_2, in one second

$$= \frac{KA(\theta_1 - \theta_2)}{d_1} + \frac{KA(\theta_3 - \theta_4)}{d_2} \qquad \text{...(ii)}$$

Equating (i) and (ii)

$$\frac{EI}{4.2} = \frac{KA(\theta_1 - \theta_2)}{d_1} + \frac{KA(\theta_3 - \theta_4)}{d_2} \qquad \text{...(iii)}$$

The value of K is calculated from equation (iii).

3.29 LEE AND CHARLTON'S METHOD FOR BAD CONDUCTORS

The apparatus consists of a cylindrical steam chamber A, the specimen disc D and brass or copper block C. The whole apparatus is suspended from the stand (Fig. 3.27). T_1 and T_2 are the thermometers used to determine the temperature after the steady state reached.

Steam is passed through the chamber and the readings of the thermometers T_1 and T_2 are noted after the steady state is reached. The heat passing through D in one second is equal to the heat radiated by the exposed surface of C in one second.

$$\frac{KA(\theta_1 - \theta_2)}{d} = ms\frac{d\theta}{dt}\left[\frac{A+S}{2A+S}\right] \qquad \text{...(i)}$$

Here $\left[\frac{A+S}{2A+S}\right]$ is the fraction of the total area exposed to the surroundings.

Here A is the area of cross-section of D and C. S is the area of the curved surface C, $\frac{d\theta}{dt}$ is the rate of cooling at temperature θ_2, m is the mass and s is the specific heat of C.

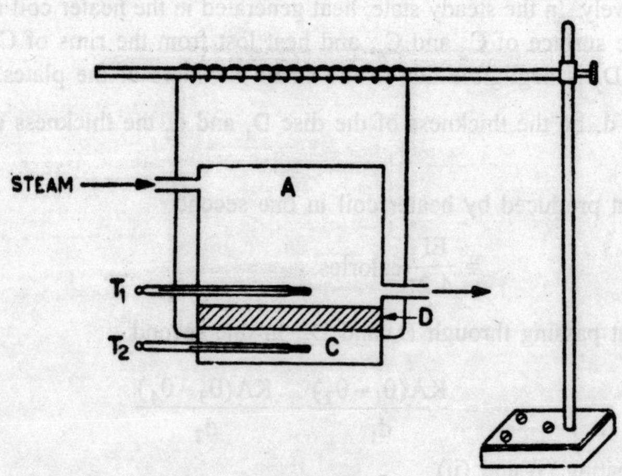

Fig. 3.27

To find $\frac{d\theta}{dt}$, the disc D is removed and C makes contact with the steam chamber. C is removed when its temperature is about 10°C higher than θ_2. It is placed over two knife edges and its temperature is observed after equal intervals of time (say one minute). A graph is drawn between temperature and time. From the graph, the value of $\frac{d\theta}{dt}$ at temperature θ_2 is found.

From equation (i), K can be calculated.

3.30 LEE'S METHOD FOR LIQUIDS

To determine the thermal conductivity of a liquid, convection current should be determined. To overcome this difficulty, heat should flow from the upper surface to the lower surface of the liquid.

Lee designed an apparatus shown in Fig. 3.28. C_1, C_2, C_3 and C_4 are the copper plates of uniform thickness. H is the heater coil. A glass plate is pressed between C_2 and C_3. The thermocouples T_1, T_2, T_3 and T_4 are used to determine the temperatures. The liquid is contained in a cylindrical tube having an ebonite ring.

A steady current of I amperes is passed through the heater coil H. After the steady state is reached, the temperatures shown by the thermocouples are noted. Let these temperatures be θ_1, θ_2, $\theta_2\theta_3$ and θ_4.

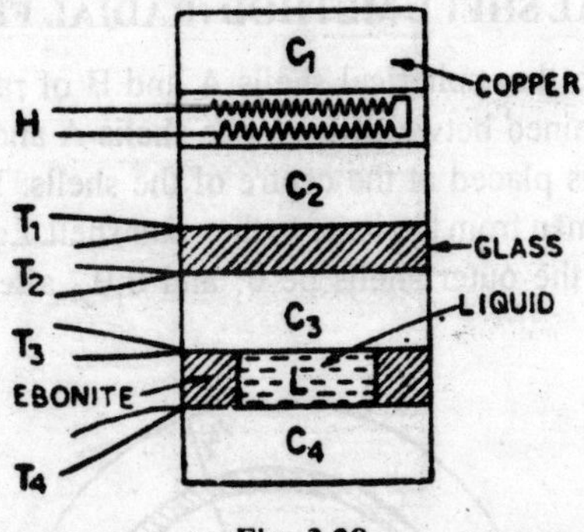

Fig. 3.28

Area of cross-section of the glass plate = A

Thickness of the glass plate = d

Area of cross-section of the liquid surface = A_1

Area of cross-section of the ebonite surface = A_2

Thickness of the liquid = d_1

Thickness of the ebonite ring = d_1

Thermal conductivity of glass = K

Thermal conductivity of liquid = K_1

Thermal conductivity of ebonite = K_2

Heat flowing per second through the glass plate

$$= \frac{KA(\theta_1 - \theta_2)}{d} \quad \text{...(i)}$$

Heat flowing per second through the liquid and the ebonite ring

$$= \frac{K_1A_1(\theta_3 - \theta_4)}{d_1} + \frac{K_2A_2(\theta_3 - \theta_4)}{d_1} \quad \text{...(ii)}$$

Equating (i) and (ii)

$$\therefore \quad = \frac{KA(\theta_1 - \theta_2)}{d} = \frac{K_1A_1(\theta_3 - \theta_4)}{d_1} + \frac{K_2A_2(\theta_3 - \theta_4)}{d_1} \quad \text{...(iii)}$$

The value of K, for the liquid can be calculated from equation (iii). However, there is small error due to the heat lost by the sides of C_1, C_2, C_3 C_4, the glass plate and the ebonite ring.

3.31 SPHERICAL SHELL METHOD (RADIAL FLOW OF HEAT)

Consider two thin spherical shells A and B of radii r_1 and r_2. The specimen is contained between these two shells A and B (Fig. 3.29). A heating element is placed at the centre of the shells. Heat is conducted through the specimen from the inner to the outer shell. Let the temperatures of the inner and the outer shells be θ_1 and $\theta_1\theta_2$, after the steady state is reached.

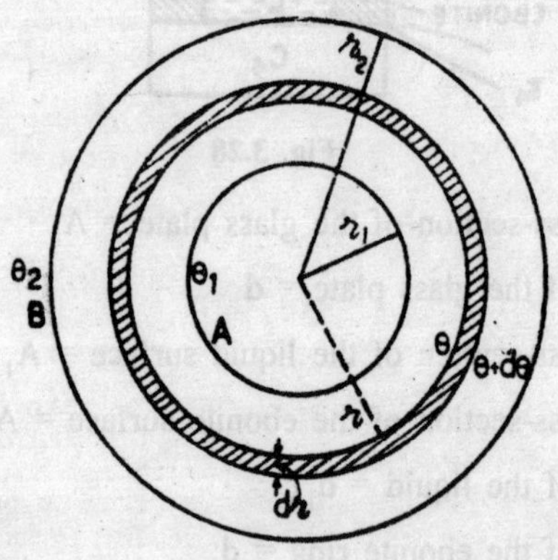

Fig. 3.29

Consider an imaginary shell of radius r and thickness dr having temperatures θ and θ + dθ on its inner and the outer surfaces respectively. The quantity of heat conducted per second through this shell,

$$Q = -KA\frac{d\theta}{dr} \qquad ...(i)$$

Here $\quad A = 4\pi r^2$

$$\therefore \qquad Q = -K.4\pi r^2\frac{d\theta}{dr}$$

$$\frac{dr}{r^2} = -\frac{4\pi K}{Q}d\theta$$

Integrating,

$$\int_{r_1}^{r_2} \frac{dr}{r^2} = -\frac{4\pi K}{Q} \int_{\theta_2}^{\theta_2} d\theta$$

$$\frac{1}{r_1} = \frac{1}{r_2} = \frac{4\pi K(\theta_1 - \theta_2)}{Q}$$

$$K = \frac{Q(r_2 - r_1)}{4\pi r_1 r_2 (\theta_1 - \theta_2)}$$

K is calculated from equation (ii).

3.32 CYLINDRICAL FLOW OF HEAT

Consider a cylindrical tube of length l, inner radius r_1 and outer radius r_2. After the steady state is reached, the temperature on the inner surface is θ_1 and on the outer surface it is θ_2. Here $\theta_1 > \theta_2$. Heat is conducted radially across the wall of the tube. Consider an element of thickness dr and length l at a distance r from the axis (Fig. 3.30).

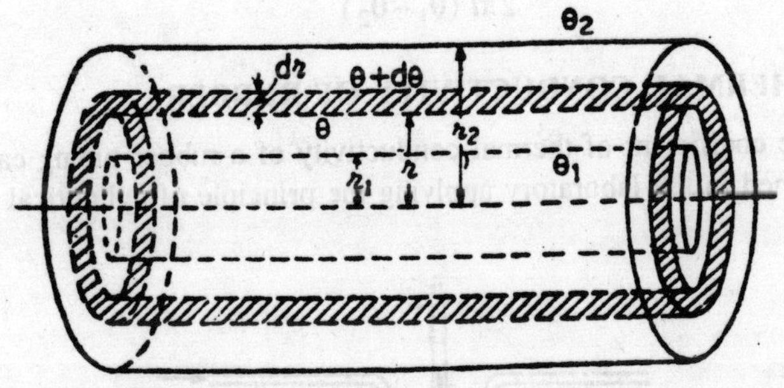

Fig. 3.30

The quantity of heat flowing per second across the element

$$Q = -KA\frac{d\theta}{dr} \qquad \text{...(i)}$$

But $\qquad A = 2\pi r l$

$\therefore \qquad Q = -K.2\pi r l\dfrac{d\theta}{dr}$

$$\frac{dr}{r^2} = -\frac{4\pi K}{Q} d\theta$$

$$Q.\ \frac{dr}{r} = -\ 2\pi r l d\theta$$

Q is constant after the steady state is reached.

Integrating equation (i)

$$Q\int_{r_1}^{r_2} \frac{dr}{r} = -2\pi K l \int_{\theta_1}^{\theta_2} d\theta$$

$$Q\left[\log_e \frac{r_2}{r_1}\right] = -2\pi K l\ [\theta_2 - \theta_1]$$

$$Q \log_e \frac{r_2}{r_1} = -2\pi K l\ [\theta_1 - \theta_2]$$

$$K = \frac{Q \log_e \frac{r_2}{r_1}}{2\pi l\ (\theta_1 - \theta_2)}$$

$$K = \frac{Q \times 2.3026 \times \log_{10} \frac{r_2}{r_1}}{2\pi l\ (\theta_1 - \theta_2)}$$

3.33 THERMAL CONDUCTIVITY OF RUBBER

The coefficient of thermal conductivity of a rubber tubing can be determined in the laboratory applying the principle of cylindrical flow of heat.

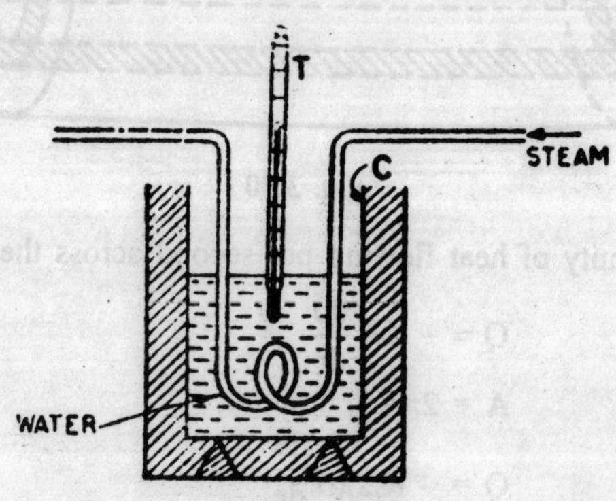

Fig. 3.31

A known quantity of water is taken in a calorimeter C. A. rubber tubing whose, inner and outer radii are r_1 and r_2, is taken and a known length (say 50 cm) of it is immersed in water as shown in Fig. 3.31. The initial temperature of water is noted. Let it be θ_3. Steam is passed through the rubber tubing for a known time t seconds (say 900 seconds). Let the final temperature of water be θ_4, (after applying radiation correction) and temperature of steam be θ_1. The average temperature on the outer surface of the rubber tubing

$$= \theta_2 = \frac{\theta_4 - \theta_3}{2}$$

Suppose, mass of water = m

Water equivalent of the calorimeter = w

Rise in temperature = $(\theta_4 - \theta_3)$

Heat gained by water = $(m + w)(\theta_4 - \theta_3)$

Quantity of heat flowing per second

$$Q = \frac{(m+w)(\theta_4 - \theta_3)}{t}$$

But
$$K = \frac{Q \times 2.3026 \times \log_{10}\left(\frac{r_2}{r_1}\right)}{2\pi l\,[\theta_1 - \theta_2]}$$

Substituting the values of θ_2 and Q

$$K = \frac{(m+w)(\theta_4 - \theta_3) \times 2.3026 \times \log_{10}\left(\frac{r_2}{r_1}\right)}{2\pi l\left(\theta_1 - \frac{\theta_3 - \theta_4}{2}\right)t}$$

Thus, K for rubber can be calculated.

3.34 THERMAL CONDUCTIVITY OF GLASS

A glass tube whose inner and outer radii are r_1, and r_2 respectively is fixed inside an outer jacket (Fig. 3.32). The apparatus is kept in the slanting position and water is allowed to flow inside the glass tube from lower to the upper end. The thermometers T_1 and T_2 are used to determine the temperature of incoming and outgoing water. The rate of flow of water can be adjusted with the help of a pinch cock.

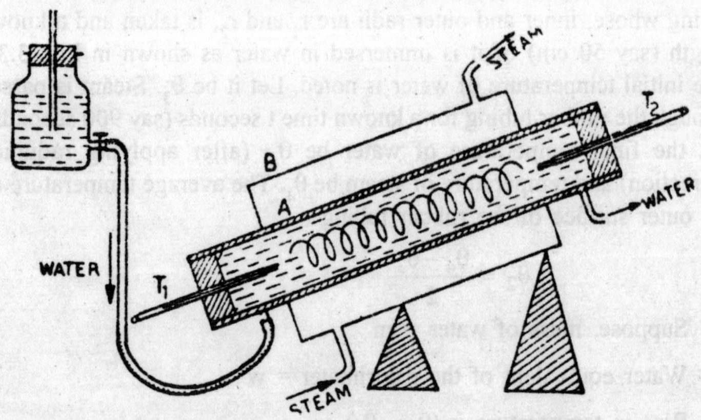

Fig. 3.32

Steam is passed inside the outer jacket and the rate of flow of water is adjusted such that the two thermometers show sufficient difference of temperature (say 5 to 10°C). When the two thermometers show constant temperature, water flowing through the glass tube for a known time t seconds (say 900 seconds) is collected in a beaker. The spiral wire inside the glass tube helps in making good contact of water with the wall of the tube.

Suppose,

Length of the tube inside the jacket = l

Temperature of steam = θ_1

Temperature of incoming water = θ_3

Temperature of outgoing water = θ_4

Mass of water collected = m

Time taken = t seconds

Quantity of heat gained by water in t seconds

$$= m(\theta_4 - \theta_3)$$

Quantity of heat that flows in one second across the wall of the tube,

$$Q = \frac{m(\theta_4 - \theta_3)}{t}$$

Average temperature of the inside of the glass tube

$$= \theta_2 = \frac{\theta_4 + \theta_3}{2}$$

But $$K = \frac{Q \times 2.3026 \times \log_{10}\left(\frac{r_2}{r_1}\right)}{2\pi l\left(\theta_1 \frac{\theta_3 + \theta_4}{2}\right)t} \quad ...(i)$$

Substituting the values of Q in equation (i)

$$K = \frac{m(\theta_4 - \theta_3) \times 2.3026 \times \log_{10}\left(\frac{r_2}{r_1}\right)}{2\pi l\left(\theta_1 - \frac{\theta_3 + \theta_4}{2}\right)t} \quad ...(ii)$$

Hence, K can be determined.

3.35 HEAT HOW THROUGH A COMPOUND WALL

Consider a compound wall (or a slab) made of two materials A and B of thickness d_1 and d_2 (Fig. 3.33). Let K_1 and K_2 be the co-efficients of thermal conductivity of the two materials θ_1 and θ_2 are the temperatures of the end faces ($\theta_1 > \theta_2$) and 6 is the temperature of the surface in contact. After the steady state is reached, the heat flowing per second (Q) across any cross-section is the same.

For the material A,

$$Q = \frac{K_1 A(\theta_1 - \theta)}{d_1}$$

For the material B,

$$Q = \frac{K_2 A(\theta - \theta_2)}{d_2}$$

From equations (i) and (ii),

$$\frac{K_1 A(\theta_1 - \theta)}{d_1} = \frac{K_2 A(\theta - \theta_2)}{d_2}$$

$$\theta = \frac{\frac{K_1\theta_1}{d_1} + \frac{K_2\theta_2}{d_2}}{\frac{K_1}{d_1} + \frac{K_2}{d_2}}$$

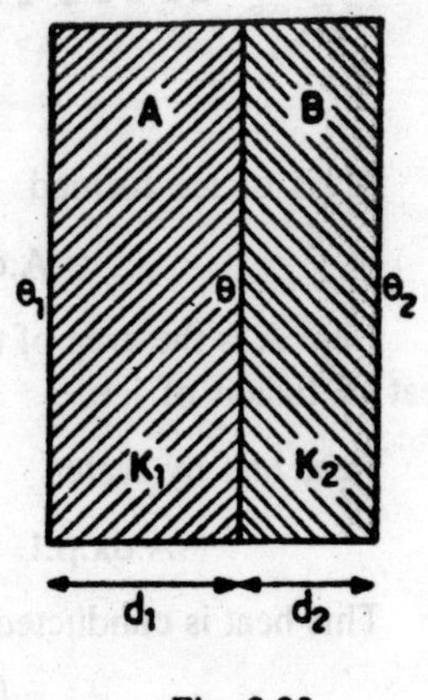

Fig. 3.33

Substituting the value of 6 in equation (i),

$$Q = \frac{A(\theta_1 - \theta_2)}{\frac{d_1}{K_1} + \frac{d_2}{K_2}} \quad ...(iii)$$

In general, for any number of walls or slabs,

$$Q = \frac{A(\theta_1 - \theta_2)}{\Sigma\left(\frac{d}{K}\right)} \quad ...(iv)$$

3.36 ACCRETION OF ICE ON PONDS

Consider a layer of ice x cm thick on the surface of a pond. Let the temperature of air over the surface of ice be θ°C and that of water below ice 0°C. Suppose a thickness dx of ice is formed in time dt (Fig. 3.34).

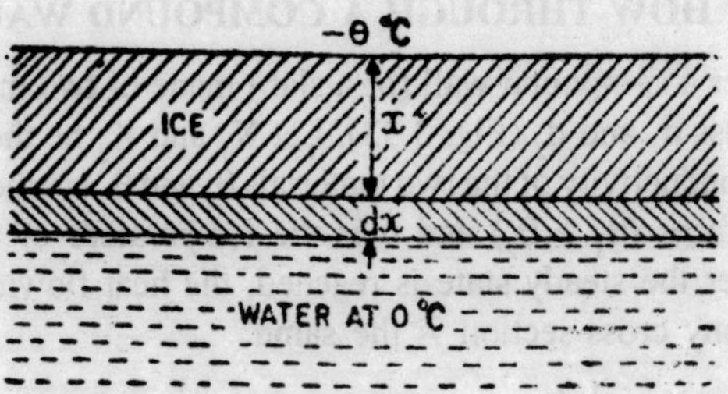

Fig. 3.34

Mass of ice formed

$$= A.dx.\rho$$

Here A is the area of the pond, ρ the density of ice and L the latent heat of fusion of ice.

Heat lost by water

$$= A.dx.\rho.L \text{ calories} \quad ...(i)$$

This heat is conducted across a layer of ice of thickness x upwards.

$$\text{Heat conducted } KA\frac{\theta}{x}dt \text{ calories} \quad ...(ii)$$

Equating (i) and (ii)

$$KA\frac{\theta}{x}dt = Adx.\rho L$$

$$\frac{dx}{dt} = \frac{K\theta}{\rho Lx} \qquad ...(iii)$$

$\frac{dx}{dt}$ represents the rate of growth of the thickness of ice.

$$dt = \left(\frac{\rho L}{K\theta}\right)xdx$$

Total time taken by the layer of ice to increase in thickness by x

$$\int dt = \frac{\rho L}{K.\theta}\int x.dx$$

$$t = \frac{\rho L}{K.\theta}\frac{x^2}{2} + \text{constant}$$

When $t = 0, x = 0$

Hence the constant is zero.

$$t = \frac{\rho L}{2K\theta}x^2 \qquad ...(iv)$$

3.37 WIEDEMANN-FRANZ LAW

In 1853, Wiedemann and Franz, based on experimental results, put forward a law relating to the thermal conductivity and electrical conductivity of metals. According to this law t.

The ratio of the thermal and electrical conductivities is the same for all metals at the same temperature. Moreover, the ratio is directly proportional to the absolute temperature of the metal.

Let K and a be thermal and electrical conductivities of a metal at a temperature T degrees absolute,

$$\frac{K}{\sigma} \propto T$$

or $$\frac{K}{\sigma T} = \text{constant for all metals.}$$

This law holds good for a large number of metals between +100°C and −100°C. At low temperatures the ratio K/σ decreases and the value tends to be zero at absolute zero. As the temperature of the metal is decreased, the thermal and electrical conductivities of the metal increase. But the increase in the electrical conductivity is higher and its value tends to infinity at absolute zero.

This corresponds to the super-conducting state of the metal.

Drude explained that thermal and electrical conductivities are due to free electrons in metals. He derived the expression

$$\frac{K}{\sigma T} = \frac{3R^2}{JN^2e^2} = \text{constant},$$

where R is the universal gas constant, N is the number of molecules in one gram molecule J is the Joule's mechanical equivalent of heat and e is the charge on the electron.

Values of $\frac{K}{\sigma T} \times 10^6$.

Metal	–100°C	–50°C	°C	18°C
Aluminium	1.81	1.98	2.09	2.13
Copper	2.17	2.26	2.30	2.32
Iron	2.98	2.93	2.97	2.99
Lead	2.54	2.52	2.53	2.51
Silver	2.29	2.36	2.33	2.33
Zinc	2.39	2.40	2.45	2.43

3.38 PRACTICAL APPLICATIONS OF CONDUCTION OF HEAT

Metals are good conductors of heat and wood, felt, brick, glass, granite, cotton, wool, cork, ebonite, rubber are bad conductors of heat.

(1) Sauce pans, hot water buckets, kettles and other utensils are made of metal. They are provided with wooden or ebonite handles so that heat from the utensil is not conducted to the hand.

(2) Ice box has a double wall made of tin or iron. The space in between the two walls is packed with cork or felt. This is done because cork and felt are poor conductors of heat and prevent the flow of the outside heat into the box.

(3) Thick brick walls are used in the construction of a cold storage. Brick is a bad conductor of heat and does not allow outside heat to flow inside the cold storage.

(4) Woollen clothes have fine pores filled with air. Air and wool are bad conductors of heat. Heat from the body does not flow outside to the atmosphere. Therefore woollen clothes keep the body warm in winter.

(5) Two shirts keep the body warmer than a single shirt of the same material and double the thickness. Between the two shirts a fine layer of air acts as a bad conductor and does not allow the heat from the body to flow out to the surroundings.

(6) Quilts and bed clothings filled with cotton are used in winter. The air layers in the pores of the cotton are bad conductors of heat. Therefore the flow of heat to outside a prevented.

(7) A steel blade appears colder than a wooden handle in winter. Steel is a good conductor of heat. As soon as a person touches the blade, heat flows from the hand to the blade. Therefore it appears colder. Since wood is a bad conductor of heat, heat does not flow from the hand to the handle.

(8) In cold countries, the windows are provided with double doors. Air in between the two doors forms a non-conducting layer and does not allow heat to flow from inside to outside.

In hot countries also, double door windows are used. Heat does not flow from outside to inside because air forms a non-conducting layer in between the two doors.

(9) When a stopper fitted tightly to the bottle is to be removed the neck is gently heated. Glass is a poor conductor of heat. Neck expands but heat is not conducted to the stopper. The stopper can be removed easily.

3.39 DAVY'S SAFETY LAMP

It also is based on the principle of conduction of heat. A wire gauze is placed over a bunsen burner and the gas is lit above the wire gauze (Fig. 3.35). The gas comes out of the wire gauze. A flame appears at the top surface of the wire gauze. The gas below the wire gauze does not get sufficient heat for ignition. The wire gauze conducts away the heat of the same above it and the temperature at the lower surface does not reach the ignition temperature.

In a Davy's safety lamp, the cylindrical metal gauze of high thermal conductivity surrounds the flame (Fig. 3.36). When this lamp is taken inside a mine, even if explosive gases are present, they do not get ignited because the wire gauze conducts away the heat of the flame. The temperature outside the wire gauze remains lower than the ignition temperature of the gases. In the absence of the wire gauze the gases outside can explode.

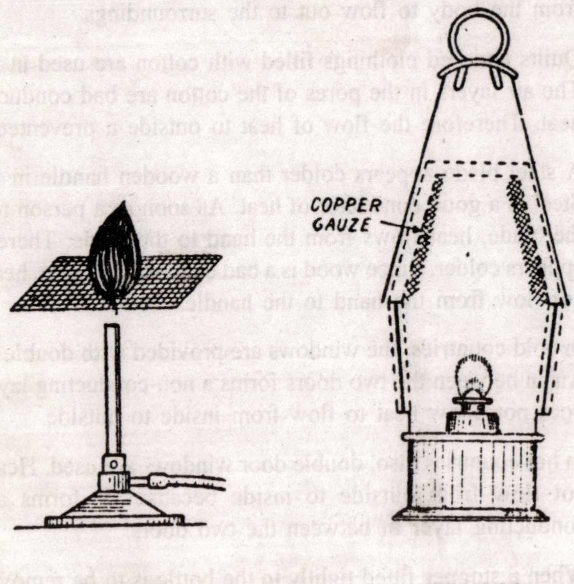

Fig. 3.35 **Fig. 3.36**

As soon as a miner notices the presence of the explosive gases from the colour of the flame, he gives a warning signal. The work is stopped and the mine is thoroughly ventilated.

3.40 CONVECTION

Experiment 1:

Take a flask containing water. Add a large crystal of $KMnO_4$. Heat the flask (Fig. 3.37). Coloured streaks of water rise up and move to the sides. Water at the bottom gets heated and moves up. Water from the sides reaches the bottom, gets heated and rises up. The process continues. This phenomenon is called convection. Every molecule of water comes to the hot point, takes heat and moves up.

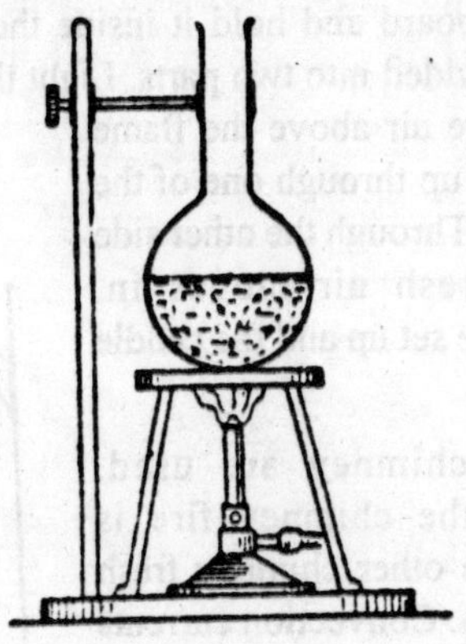

Fig. 3.37

Experiment 2:

Take a rectangular glass tube and fill it with water (Fig. 3.38). Pour a little charcoal powder in the tube. Heat the side tube gently. The convection currents are set up due to the movement of the heated water molecules. The direction of movement of water molecules is indicated by the movement of the charcoal particles.

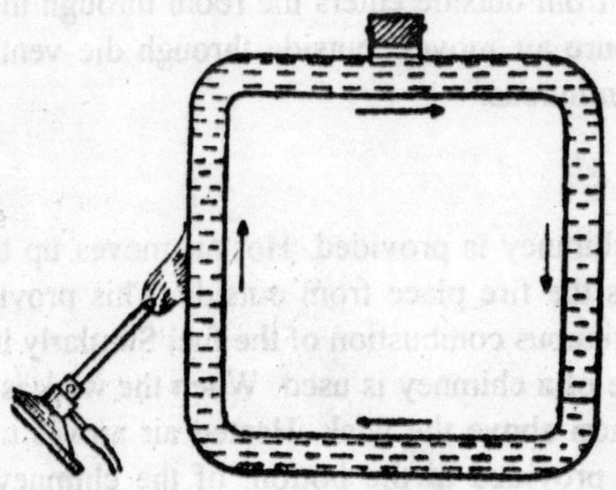

Fig. 3.38

Experiments 3:

Take a candle and fix it at the bottom of a cylinder (Fig. 3.39). Light the candle. The flame becomes weaker and weaker and finally gets extinguished. Air in the cylinder gets heated and is pushed out. There is no fresh supply of air for the burning of the candle.

Now, take a cardboard and hold it inside the cylinder. The space above the candle is divided into two parts. Light the candle. The candle continues to burn. Here air above the flame gets heated and moves up through one of the sides of the cardboard. Through the other side of the cardboard fresh air moves in. Convection currents are set up and the candle continues to burn.

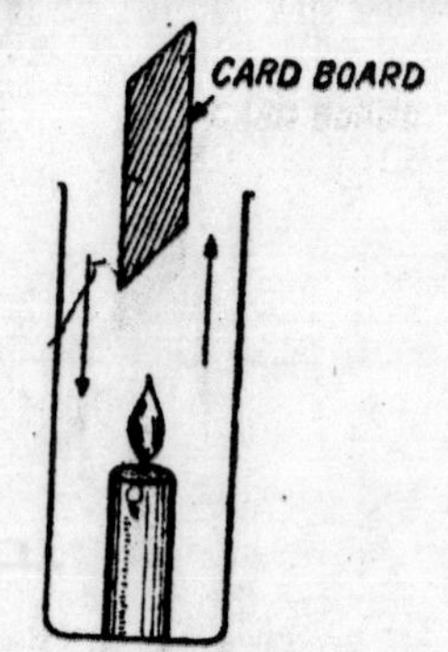

Fig. 3.39

In mines, two chimney are used. Underneath one of the chimney fire is produced. Through the other chimney fresh air enters into the mines. Convection currents are produced. This provides proper ventilation in the mine.

3.41 APPLICATION OF CONVECTION

1. Ventilation

Rooms are provided with ventilators near the ceiling. Air in the room gets warmer due to respiration of people in the room. Warm air containing more of CO_2 and water vapour has less density and moves upwards. Fresh air from outside enters the room through the doors and windows. The impure air moves, outside through die ventilators. The phenomenon is continuous.

2. Chimneys

In a kitchen, chimney is provided. Hot air moves up the chimney and fresh air enters the fire place from outside. This provides enough oxygen for the continuous combustion of the fuel Similarly in off banns, oil stoves, principle of a chimney is used. When the wick is lighted, air and kerosene oil burn above the wick. Heated air moves up and fresh air from the holes provided at the bottom of the chimney enters the inside. The phenomenon is continuous and the convection currents set up help in supplying enough oxygen for combustion. If holes are not provided, the flame gets extinguished quickly.

In a factory, a high chimney is provided. The rate of circulation of air depends on the difference in pressure between the two ends of the chimney. Higher the chimney, greater is the rate of circulation of air. Due to this reason, sufficient supply of oxygen is provided and the complete combustion of thc fuel takes place.

3. Trade Winds

The surface of the earth gets heated more at the equator than at me poles. Warm air at the equator moves up and cold air from the poles moves towards the equator. In the northern hemisphere, if is coming from the north and due to the rotation of the earth from west to cast, the wind appears to come from north-east. In the southern hemisphere, the wind appears to be from south-west. These winds are called trade winds because they were used by traders for sailing their vessels in ancient days.

Monsoons are also based on the principle of convection.

4. Ocean Currents

Water in the ocean at the equator gets heated more than at the polo. Heated water at the equator expands and the level of water rises near the equator. This hot water flows towards the poles. This current is called hot current or gulf streams. The cold water from the poles moves towards the equator beneath the ocean and this current is called the under current.

5. Land and Sea Breezes

Near the sea, water becomes less warm than the land during day time. The heated air on the land surface moves up. Air from the surface of the sea moves towards the sea shore. This is called sea breeze. Convection currents are set up During night when the land becomes cooler than water, the air over the surface of water is warmer and moves upward. Air from the land moves towards the sea.This is called land breeze.

3.42 CENTRAL HEATING SYSTEM

In cold countries, the temperature in Winter falls below 0°C. The rooms of a building are kept warm by a central heating system based on the principle of convection.

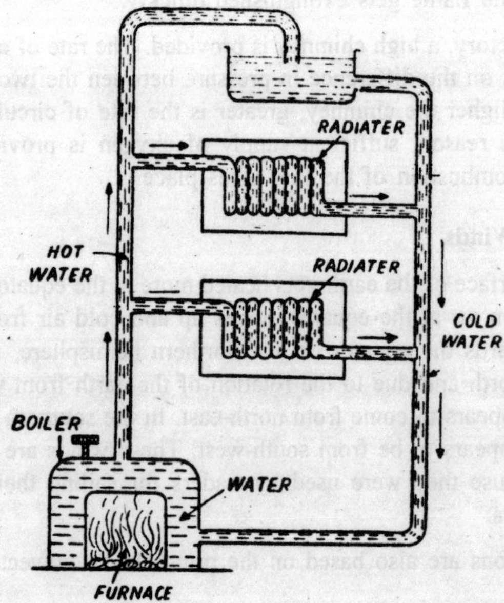

Fig. 3.40

Water is heated in the boiler at the basement of the building. The rooms are fitted with pipes having radiators fixed on the walls. Hot water from the boiler rises up. It passes through the radiators of different rooms. Radiators get heated and radiate heat to the room. After losing heat to the radiator, water becomes cold and returns back to the boiler as shown in Fig. 3.40. Hot water also reaches the cold water tank at the top of the building. The hot water flow a from the boiler to the cold water tank and the cold water flows from the tank to the boiler. Convection currents are set up and the building is kept warm continuously at a constant temperature.

3.43 CHANGE OF PRESSURE WITH HEIGHT

The pressure of the atmosphere decreases with increase in height from sea level. The density of air is not the same at all levels. If the density of air were the same at all levels, the atmospheric pressure at sea level would correspond to a vertical column of 8 km of air only. But it has been found that air exists even at a height of 200km.

Consider that the pressure of air at a height h = P. For an increase in height dh, the decrease in pressure is dP (Fig. 3.41).

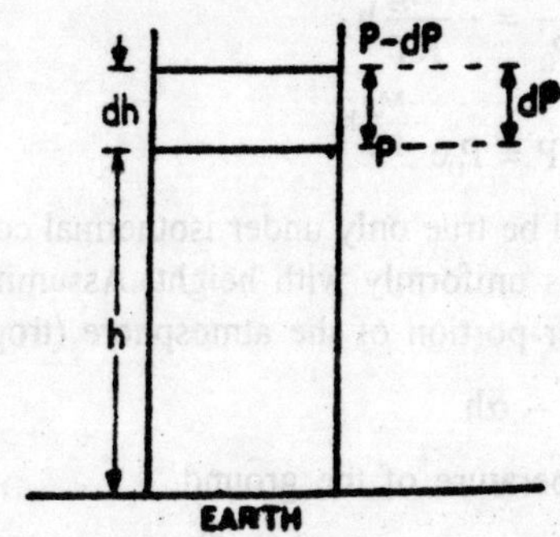

Fig. 3.41

$$dP = -(dh)\rho g \qquad ...(i)$$

[–ve sign shows that the pressure decreases with increase in height]

Let M be the molecular weight of the gas and V its volume.

Then, $\rho = \dfrac{M}{V}$

and $dP = -(dh)\dfrac{M}{V}g$

But for a perfect dry air

$$PV = RT$$

or $V = \dfrac{RT}{P}$

$\therefore$ $dP = -(dh)\dfrac{M.Pg}{RT}$

$$\frac{dP}{P} = -\frac{Mg}{RT}(dh) \qquad ...(ii)$$

Integrating

$$\log P = -\frac{Mg}{RT}h + C \quad \text{[where C is a constant]}$$

When $h = 0$,

$P = P_0$

$\therefore$ $C = \log P_0$

and $$\log P = -\frac{Mg}{RT}h + \log P_0$$

$$\log_e \frac{P}{P_0} = -\frac{Mg}{RT}h \quad \text{...(iii)}$$

$$P = P_0 e^{-\frac{Mg}{RT}h} \quad \text{...(iv)}$$

Equation (ii) will be true only under isothermal conditions. But air temperature decreases uniformly with height. Assuming that the lapse rate is α in the lower portion of the atmosphere (troposphere),

$$T = T_e - \alpha h$$

where T_0 is the temperature of the ground.

Substituting this value in equation (ii)

$$\therefore \quad \frac{dP}{P} = -\frac{Mg}{R}\frac{dh}{(T_0 - \alpha h)}$$

$$\log P = \frac{Mg}{R\alpha}\log(T - \alpha h) + K \quad \text{...(v)}$$

$$h = 0, \qquad P = P_0$$

$$\therefore \quad \log P_0 = \frac{Mg}{R\alpha}\log T_0 + K$$

or $$K = \log P_0 - \frac{Mg}{R\alpha}\log T_0$$

Substituting this value of K in equation (v)

$$\log P = \frac{Mg}{R\alpha}\log(T_0 - \alpha h) + \log P_0 - \frac{Mg}{R\alpha}\log T_0$$

$$\log\frac{P}{P_0} = \frac{Mg}{R\alpha}\log\left(\frac{T_0 - \alpha h}{T_0}\right) \quad \text{...(vi)}$$

$$P = P_0 e^{\frac{Mg}{R\alpha}\log\left(\frac{T_0 - \alpha h}{T_0}\right)} \quad \text{...(vii)}$$

3.44 CONVECTIVE EQUILIBRIUM OF THE ATMOSPHERE

The atmospheric temperature decreases with altitude upto a height of about 20 km. The rate of fall of temperature is about 5oC per km height and is known as the lapse rate. The atmosphere is divided into two regions.

1. *Troposphere or the convective zone :* It is the region in which the temperature falls with increase in height from the ground.
2. *Stratosphere or advective zone*: It is the region in which the decrease in temperature does not take place with increase in height.

The layer that separates these two regions is known as *tropopause*. The height of tropopause is different at different places on the earth. It is about 14 km at the equator and about 10 km at the poles.

The fall in temperature with altitude in the troposphere is due to convection. When the heat radiations from the sun fall on the surface of the earth, it gets heated. The atmospheric air surrounding the earth gets heated by conduction and radiation. The heated air moves up and convection currents are set up. The heated air that moves up from the regions of higher to lower pressure is adiabatically cooled. Similarly when the colder air moves down from lower to higher pressure regions it is adiabatically heated. Thus, a *convective* equilibrium is set up and there is a gradual fall of temperature with increase in height.

The earth gets heated up due to the heat radiations from the sun. The layers of the atmosphere close to the ground get heated up and convection currents start in the atmosphere. As the warm air rises up into the region of low pressure, it *expands adiabatically*. For an atmosphere, in convective equilibrium, there is a vertical fall of temperature from the earth's surface. The *dry adiabatic lapse rate* is calculated as follows:

The relation between pressure and temperature for an adiabatic process is

$$\frac{P^{\gamma-1}}{T^{\gamma}} = \text{const.}$$

or

$$P^{\gamma-1}\,T^{-\gamma} = \text{const.}$$

Differentiating

$$(\gamma - 1)\,P^{\gamma-2}dP.T^{-\gamma} + P^{\gamma-1}(-\gamma)T^{-\gamma-1}dT = 0$$

Dividing by $P^{\gamma-1}T^{-\gamma}$

$$(\gamma - 1)\frac{dP}{P} + (-\gamma)\frac{dT}{T} = 0$$

$$\frac{dP}{P} = \frac{\gamma}{\gamma - 1}\frac{dT}{T} \qquad \text{...(i)}$$

For an increase in height dh, the decrease in pressure is dP

$$\therefore \quad dP = -\rho g\, dh = -\frac{M}{V} g\, dh$$

But $\quad PV = RT$

or $\quad V = \dfrac{RT}{P}$

$$\therefore \quad dP = -\frac{MP}{RT} g\, dh$$

$$\frac{dP}{P} = -\frac{Mg}{RT} dh \quad \text{...(ii)}$$

Equating (i) and (ii)

$$\left(\frac{\gamma}{\gamma - 1}\right)\frac{dT}{T} = -\frac{Mg}{RT} dh$$

$$\frac{dT}{dh} = -\frac{Mg}{R}\left(\frac{\gamma - 1}{\gamma}\right) \quad \text{...(iii)}$$

This equation holds good only for perfect dry air. The lapse rate $\dfrac{dT}{dh}$ can be calculated from equation (iii).

Substituting the values,

$M = 29$ g

$g = 981$ cm/s^2

$R = 8.31 \times 10^7$ ergs per gram mol per K,

and $\quad \gamma = 1.4$

$$\frac{dT}{dh} = -\frac{2.9 \times 981}{8.31 \times 10^7}\left(\frac{1.40 - 1}{1.40}\right)$$

$= 10^{-4}$ K per cm

$= 10^{-4}$ °C per cm.

Thus, the lapse rate for a height of 1 km $= 10^{-4} \times 10^5 = 10$°C. But it has been found that the average lapse rate is lower than this value. Under average conditions the lapse rate has been found to be between 5°C and 6-5°C per km.

However, at night, when the earth is radiating heat to the atmosphere, the layers of air nearer the surface may cool to a temperature lower than that of the upper layers. Then a negative lapse rate is set up.

3.45 DIFFERENTIAL AIR THERMOSCOPE

A differential air thermoscope consists of two glass bulbs A and B connected by a narrow glass tubing. The U-tube contains sulphuric acid. The bulbs A and B contain air and when they are at the same temperature, there is no difference in level of the liquid in the two limbs. The bulb A is coated with lamp black so that it can completely absorb the whole of the heat radiations falling on it (Fig. 3.42).

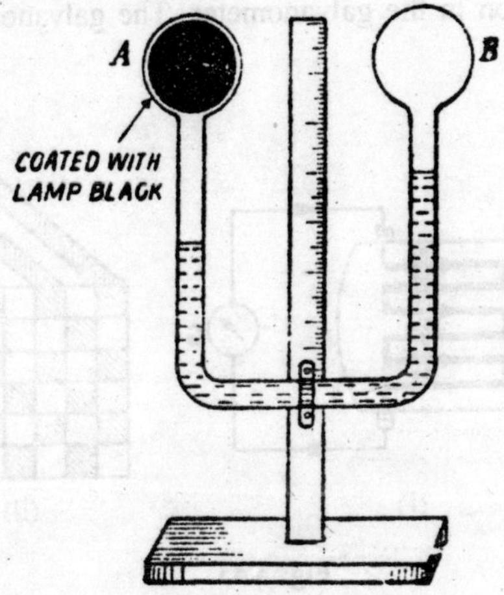

Fig. 3.42

When A is exposed to heat radiations, it absorbs heat energy, Air inside the bulb A expands and there is difference in level of the liquid in the two limbs. This thermoscope is very sensitive and can detect heat radiations of even feeble intensity (*i.e.*, from a candle at a large distance).

3.46 THERMOPILE

The thermopile originally designed by *Melloni* consisted of a number of thermocouples joined in series. The thermocouples are made of bismuth and antimony. The junctions on one side are coated with lamp black and are exposed to heat radiations The junctions on the other side are kept cold. The two ends of the thermopile are connected to a sensitive galvanometer (Fig. 3.43).

The deflection in the galvanometer measures the intensity of heat radiations and the scale can be calibrated.

The improved form of a thermopile, which is commonly used, consists of a large number of thermopiles. The bars of antimony and bismuth are arranged in the form of a cube. The junctions are soldered but the bars are insulated from each other with mica strips (Fig. 3.43), All the junctions on one side are coated with lamp-black. The galvanometer is connected to the points A and B. When not in use, the hot face is covered with a metallic cup. Such a thermopile can detect heat radiations from a candle at a distance of 100 metres. When the thermo-electric current produced in the thermopile flows through the galvanometer, there is a deflection in the galvanometer. The galvanometer scale can be calibrated.

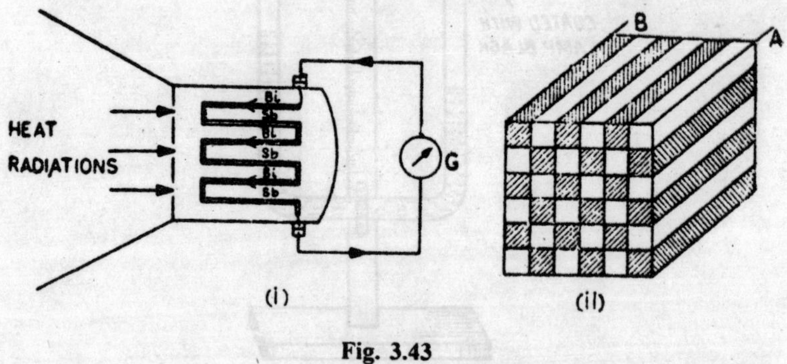

Fig. 3.43

The *linear thermopile* is used in research for the study of infrared radiations. It consists of only a few pairs of thermocouples connected to a sensitive galvanometer. The hot junctions are arranged in a line.

The thermopiles can be made sensitive by mounting the thermopile in vacuum and using very short connecting wires. The chief drawback of a thermopile consists in the comparative slowness of its indication and also the long time it takes to return to its zero reading.

3.47 PROPERTIES OF THERMAL RADIATION

Heat radiations have properties similar to light radiations. They are electromagnetic waves of longer wavelength. The range of wavelengths of heat radiations is from 3 cm to 8×10^{-5} cm.

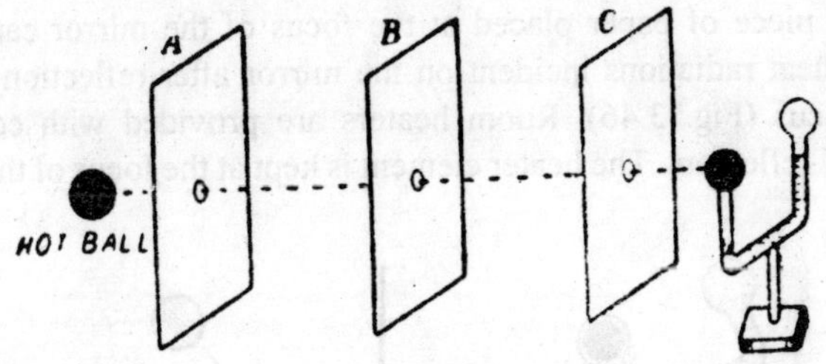

Fig. 3.44

(1) Heat radiations travel in straight lines.

Take three similar cardboards having holes at the same height. Place a red hot ball in front of the hole of the first cardboard. Bring the blackened bulb of the differential air thermoscope near the hole of the third cardboard (Fig. 3.44). When the hot ball, the three holes and the bulb of the thermoscope are in the same straight line, maximum difference in level is noted. When one of the cardboards is displaced, the thermoscope is not affected. This shows that heat radiations travel in straight lines.

(2) Heat radiations travel with the velocity of light.

When the sun rises, heat and light are received simultaneously on the earth. During the solar eclipse heat and light are cut off at the same moment. When the eclipse is over, heat and light reach the earth simultaneously. It shows that the heat radiations travel with the velocity of light equal to 3×10^8 m/s.

(3) Heat radiations obey inverse square law

Take a red hot ball. Place the thermopile at a fixed distance from the ball. Note the reading in the galvanometer. Now, place the thermopile at double the distance from the ball. It is found that the deflection of the galvanometer is one-fourth of the deflection in the first case. It shows that heat radiations obey inverse square law.

(4) Heat radiations obey the laws of reflection.

Take two tubes A and B hinged at C (Fig. 3.45). Keep the metal ball at the mouth of A. Adjust the position of B to get the maximum difference in level in the thermoscope. It is found that the difference in level is maximum when the two tubes are equally inclined to the normal at C. It means $\angle i = \angle r$.

Take a concave metallic reflecting surface. Direct it towards the sun. A small piece of paper placed at the focus of the mirror catches fire. All the heat radiations incident on the mirror after reflection converge at its focus (Fig. 3.46). Room heaters are provided with concave or elliptical reflectors. The heater element is kept at the focus of the mirrors.

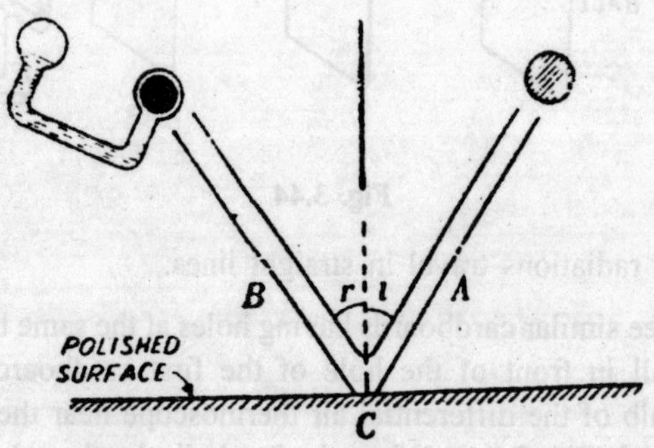

Fig. 3.45

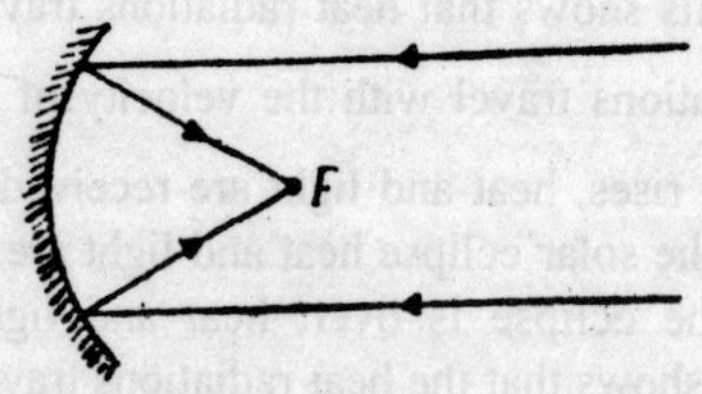

Fig. 3.46

(5) Heat radiations obey the laws of refraction

Take a convex lens and keep a small paper at its focus. When the lens is directed towards the sun's rays, the paper catches fire. The parallel heat radiations from the sun after refraction from the lens converge at its focus.

(6) Heat radiations get diffused when they are incident on a rough and unpolished surface.

(7) Heat radiations can travel through vacuum.

(8) Heat radiations do not affect the medium through which they pass. When sun's rays are allowed to pass through a convex lens, they converge at the focus of the lens. The lens does not get heated but a paper at the focus catches fire.

3.48 APPLICATIONS OF HEAT RADIATIONS

(1) White clothes are preferred in summer and dark coloured clothes in winter.

When heat radiations fall on white clothes, they are reflected back. No heat is absorbed by the clothes and a person does not get heat from outside in summer. Dark clothes in winter will absorb the heat radiations falling on them and keep the body warm.

(2) Cooking utensils are blackened at the bottom and polished on the upper surface. Black surface will absorb the whole of the heat from the furnace and the upper surface will not allow the heat inside the utensil to flow out.

(3) Hot water pipes and radiators used in rooms are painted black so that they can radiate maximum amount of heat to the room. The same pipes outside the rooms are painted white so that they do not lose heat to the surroundings.

(4) The thermocouple junction exposed to heat is blackened to absorb maximum quantity of heat.

(5) Polished reflectors are used in electric heaters to reflect maximum heat in the room.

3.49 THERMOS FLASK

A thermos flask consists of a double walled glass bottle. The space in between the two bottles is evacuated and scaled. The inner surface of B and the outer surface of A are highly polished (*i.e.*, silvered) (Fig. 3.47).

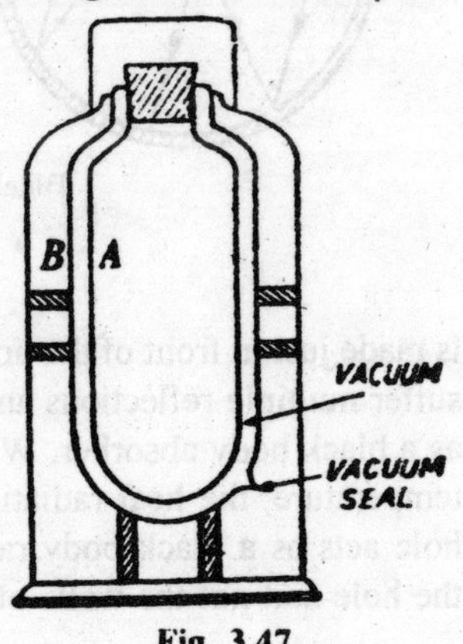

Fig. 3.47

When a hot liquid is kept in the bottle, it remains hot for a long time. The outer polished surface of A does not radiate heat. Even if a little quantity of heat is radiated by A, most of it is reflected back by the polished surface of B.

The evacuation of the space in between A and B helps in the elimination of the loss of heat by conduction and convection.

Similarly, ice kept inside the flask will not melt for a long time because heat radiations from outside cannot enter the inside of the bottle.

3.50 BLACK BODY

A perfectly black body is one which absorbs all the heat radiations (corresponding to all wavelengths) incident on it. When such a body is placed inside an isothermal enclosure, it will emit the full radiation of the enclosure after it is in equilibrium with the enclosure. These radiations are independent of the nature of the substance. Such heat radiations in a uniform temperature enclosure are known as *black body* radiations. Also the black body completely absorbs heat radiations of all wavelengths. Thus the black body also emits completely the radiations of all wavelengths at that temperature.

In practice, a perfectly black body is not available. A body showing close approximation to a perfectly black body can be constructed.

A hollow copper sphere is taken and is coated with lamp black on its inner surface (Fig. 348). A fine hole is made and a pointed projection

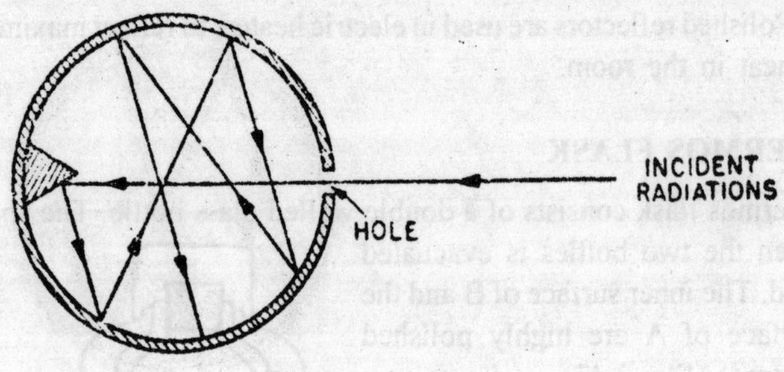

Black body absorber

Fig. 3.48

is made just in front of the hole. When the radiations enter the hole, they suffer multiple reflections and are completely absorbed. This body acts as a black body absorber. When this body is placed in a bath at a fixed temperature, the heat radiations come out of the hole (Fig. 3.49). The hole acts as a black body radiator. It should be remembered that only the hole and not the walls of the body, acts as the black body radiator.

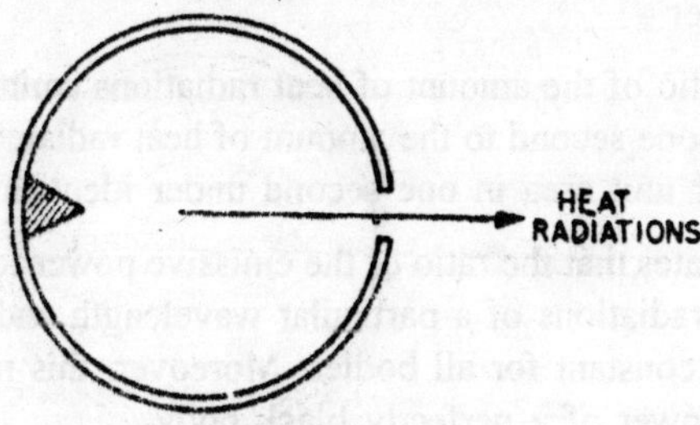

Black body emitter

Fig. 3.49

Wien also constructed a black body in the form of a cylinder. This black body is commonly used now a days.

It consists of a hollow metallic cylinder and fitted with heating coils wound around it. The inner surface of the cylinder is coated with lamp black. The cylinder is placed in concentric porcelain tubes (Fig. 3.50). The temperature is measured with the help of a thermocouple arrangement. Heat radiations emerge out of the hole.

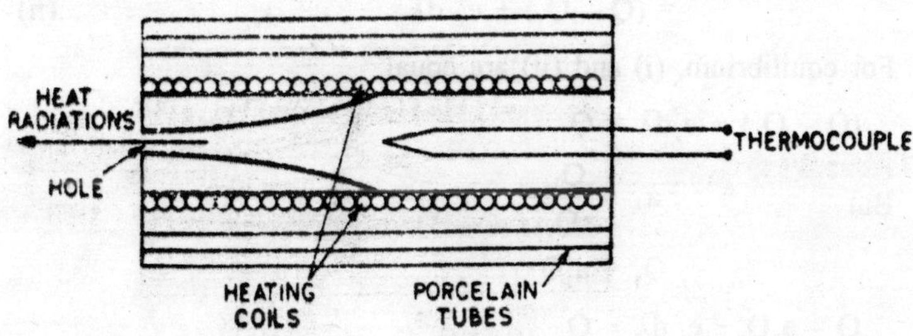

Fig. 3.50

The radiations from the inner chamber can be limited with the help of diaphragms provided on the inner side. This hole will act as a black body radiator.

3.51 KIRCHHOFF'S LAW OF HEAT RADIATION

Absorptive Power

It is the ratio of the amount of heat absorbed in a given time by the surface to the amount of heat incident on the surface in the same time.

Emissive Power

It is the ratio of the amount of heat radiations emitted by unit area of a surface in one second to the amount of heat radiated by a perfectly black body per unit area in one second under identical conditions.

The law states that the ratio of the emissive power to the absorptive power for the radiations of a particular wavelength and at a particular temperature is constant for all bodies. Moreover, this ratio is equal to the emissive power of a perfectly black body.

Suppose e_λ and a_λ are the emissive and absorptive powers of a body. Let Q be the quantity of heat radiations incident on the surface in one second, Q_1 is the quantity of heat adsorbed by the surface and Q_2 is the quantity of heat reflected by it.

As the body radiates heat, at all temperatures, the heat emitted by it in one second = $e_\lambda d\lambda$

Total heat falling on the surface = Q ...(i)

Total heat given out by the surface

$$= (Q - Q_1) + e_\lambda\, d\lambda \quad \text{...(ii)}$$

For equilibrium, (i) and (ii) are equal

$$\therefore (Q - Q_1) + e_\lambda d\lambda = Q$$

But
$$a_\lambda = \frac{Q_1}{Q}$$

$$Q_1 = a_\lambda Q$$

$$Q - a_\lambda Q + e_\lambda\, d\lambda = Q.$$

or
$$\frac{e_\lambda}{a_\lambda} = \frac{Q}{d\lambda} = \text{const.} \quad \text{...(iii)}$$

For a perfectly black body, emissive power = E_λ and absorptive power $a_\lambda = 1$.

$$\therefore \quad \frac{E_\lambda}{1} = \frac{Q}{d\lambda} \quad \text{...(iv)}$$

From equations (iii) and (iv)

$$\frac{e_\lambda}{a_\lambda} = E_\lambda = \text{const} \quad \text{...(v)}$$

Result: If (iii) is large, a_λ is also large. But

$$r_\lambda = \frac{Q - Q_1}{Q}$$

$$= 1 - a_\lambda.$$

Therefore r_λ is small. Here r_λ is the reflecting power of the surface. It means good emitters are good absorbers but bad reflectors. Dull black surfaces are good emitters and good absorbers but bad reflectors.

If e_λ is small, a_λ is also small and r_λ is large. It means bad emitters are bad absorbers but good reflectors. Highly polished surfaces are bad emitters and bad absorbers but good reflectors.

These results are verified by Ritchie's experiment.

Ritchie's Experiment

It consists of two cylindrical vessels A and B connected to the glass tubing XY (Fig. 2.51). One set of faces of A and B a blackened while the other set of faces is polished. C is another cylindrical vessel containing hot water and fixed on a stand between A and B so that the three cylinders are coaxial. One face of C is blackened while the other is polished. XY serves as a differential air thermoscope.

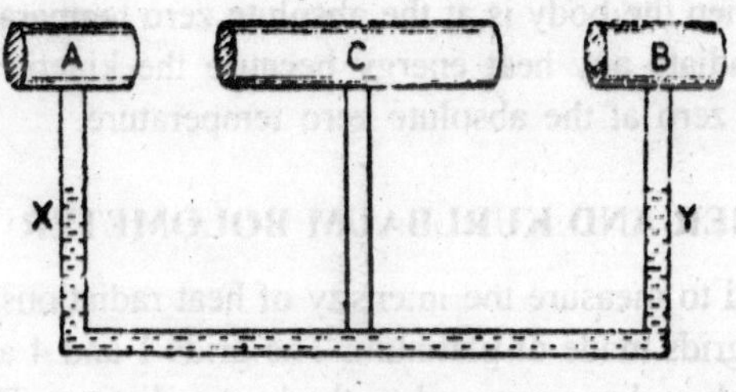

Fig. 3.51

Initially, the blackened face of C is towards the polished face of A and the polished face of C is towards the blackened face of B. The level of the liquid in the limbs X and Y will be the same. Now, turn C so that the polished face of C is towards the polished face of A. The quantity of heat emitted by the polished face of C is very small. Also the quantity of heat absorbed by the polished face of A is small. Therefore the rise in temperature of air inside A is very small. On the other hand the quantity of heat emitted by the blackened face of C is large and also the quantity of heat absorbed by the blackened face of B is large. Therefore, rise in temperature of air inside B is higher than that in A. Consequently, the level of liquid in the limb Y is lower than that in X.

Hence, good absorbers are good emitters and bad absorbers are bad emitters,

3.52 PREVOST THEORY OF HEAT EXCHANGE

According to this theory, every body radiates heat continuously at all temperatures. The quantity of heat radiated per unit area of the surface in unit time is dependent on the temperature of the body and not on the surroundings. Suppose a body A is at a higher temperature than the surroundings. It will radiate more quantity of heat to the surroundings and it will absorb less heat from the surroundings. Consequently, the temperature of the body falls. When the temperature of the body becomes equal to that of the surroundings, it is in thermal equilibrium with the surroundings. The amount of heat radiated is equal to the amount of heat absorbed by it. Similarly, when a body is at a temperature lower than that of the surroundings, it will radiate less heat and absorb more heat.

Thus, all bodies at all temperatures are in a state of dynamic thermal equilibrium when they are at the same temperature. Moreover, the amount of heat radiated from a body decreases with the fall in temperature of the body. When the body is at the absolute zero temperature (–273°C), it will not radiate any heat energy because the kinetic energy of the molecules is zero at the absolute zero temperature.

3.53 LUMMER AND KURLBAUM BOLOMETER

It is used to measure the intensity of heat radiations. It consists of four similar grids made of platinum. The grids 1 and 4 are coated with platinum black and are exposed to the heat radiations The grids 2 and 3 are shielded.

The grids are connected in the four arms of a Wheatstone's bridge as shown in Fig. 3.52. Initially the balance point is adjusted. When the heat radiations fall on 1 and 4, their electrical resistance increases and the balance point is disturbed. The deflection in the galvanometer is observed. The scale of the galvanometer can be initially calibrated with heat radiations of known intensity. This bolometer is very sensitive and can detect very small changes of temperature of the order of 10^{-4} °C.

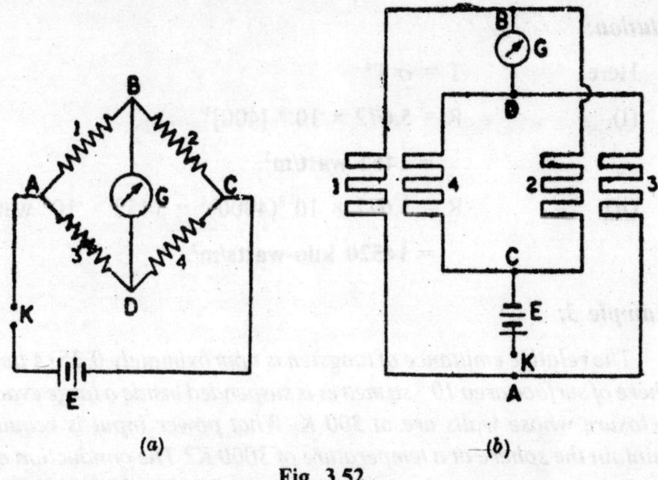

Fig. 3.52

SOLVED EXAMPLES

Example 1:

Two large closely spaced concentric spheres (both are black body radiators) are maintained, at temperatures of 200 K and 300 K respectively. The space in between the two spheres is evacuated. Calculate the net rate of energy transfer between the two spheres. (σ = 5.672 × 10^{-8} M.K.S. units).

Solution:

Here $T_1 = 300$ K

$T_2 = 200$ K

$\sigma = 5.672 \times 10^{-8}$ M.K.S. units

$R = \sigma(T_1^4 - T_2^4)$

$= 5.672 \times 10^{-8}[(300)^4 - (200)^4]$

$\mathbf{R = 368.69}$ **watts/m².**

Example 2:

Calculate the radiant emittance of a black body at a temperature of (i) 400 K (ii) 4000 K. (σ = 5.672 × 10^{-8} M.K.S. units).

Solution:

Here $T = \sigma T^4$

(i) $R = 5.672 \times 10^{-8} [400]^4$

$= \mathbf{1452\ watt/m^2.}$

(ii) $R = 5.672 \times 10^{-8}(4000)^4 = 1452 \times 10^4$ watts/m^2

$= \mathbf{14520\ kilo\text{-}watts/m^2.}$

Example 3:

The relative emittance of tungsten is approximately 0.35. A tungsten sphere of surface area 10 ^{8} sq metres is suspended inside a large evacuated enclosure whose walls are at 300 K. What power input is required to maintain the sphere at a temperature of 3000 K? The conduction of heat along the supports can be neglected. (σ = 5.672 × 10 ^{8} M.K.S. units).

Solution:

Here $R = \sigma A e[T^4 - T_0^4]$

$A = 10^{-3}$ sq meters

$e = 0.35$

$\sigma = 5.672 \times 10^{-8}$ M.K.S. units

$R = 5.672 \times 10^{-8} \times 10^{-3} \times 0.35[(3000)] - (300)^4]$

= **1609 watts**

The power input = **1609 watts.**

Example 4:

An aluminium foil of relative emittance 0.1 is placed in between two concentric spheres at temperatures 300 K and 200 K respectively. Calculate the temperature of the foil after the steady state is reached. Assume that the spheres are perfect black body radiators. Also calculate the rate of energy transfer between one of the spheres and the foil. (σ = 5.672 × 10^{-8} M.K.S. units).

Solution:

Here $T_1 = 300$ K

$T_2 = 200$ K

$$e = 0.1$$

$$\sigma = 5.672 \times 10^{-8} \text{ M.K.S. units.}$$

(i) Let x be the temperature of the foil. After the steady-state is reached,

$$e\sigma(T_1^4 - x^4) = e\sigma(x^4 - T_2^4)$$

or $$[(300)^4 - x^4] = [x - 200)^4]$$

$$x^4 = 48.5 \times 10^8$$

$$\mathbf{x = 263.8\ K}$$

(ii) $$R = e\sigma(T_1^4 - x^4)$$

$$R = 0.1 \times 5.672 \times 10^{-8}[(300)^4 - (263.8)^4]$$

$$\mathbf{R = 18.5\ watt/m^2}$$

Example 5:

Calculate the energy radiated per minute from the filament of an incandescent lamp at 2000 K, if the surface area is 5.0 × 10⁻⁵ sq metres and its relative emittance is 0.85.

Solution:

$$E = Ae\ \sigma t\ (T^4)$$

$$A = 5 \times 10^{-5} m^2$$

$$e = 0.85$$

$$\sigma = 5.672 \times 10^{-8} \text{ M.K.S. units.}$$

$$t = 60s$$

$$T = 2000\ K$$

$$E = 5 \times 10^{-6} \times 0.85 \times 5.672 \times 10^{-8} \times 60 \times (2000)^4$$

$$\mathbf{E = 2315\ joules.}$$

Example 6:

An iron furnace radiates 1.53×10^5 calories per hour through an opening of cross-section 10^{4} sq metre. If the relative emittance of the furnace is 0.80, calculate the temperature of the furnace. (Given $\sigma = 1.36 \times 10^{8}$ cal/m²-s-K⁴).

Solution:

$$E = Ae\,\sigma\, f(T^4)$$

$$E = 1.53 \times 10^5 \text{ calories}$$

$$A = 10^{-4} \text{ m}^2$$

$$e = 0.80$$

$$t = 3600 \text{ s}$$

$$T = ?$$

$$T^4 = \frac{E}{Ae\sigma t}$$

$$T = \left(\frac{E}{Ae\sigma t}\right)^{1/4}$$

$$= \left(\frac{1.53\times10^5}{10^{-4}\times0.80\times1.36\times10^{-8}\times3600}\right)^{1/4}$$

$$\mathbf{T = 2500\ K.}$$

Example 7:

Calculate the black body temperature of the sun from the following data.

Solution:

Stefan's constant $= 1.37 \times 10^{-12}$ cal/cm²/s

Solar constant $= 2.3$ cal/cm²/minute

Radius of the sun $= 7 \times 10^{10}$ cm.

Distance between the sun and the earth

$= 1.5 \times 10^{13}$cm

$$E = \frac{4\pi R^2 S}{4\pi r^2} = \left(\frac{R}{r}\right)^2 .S$$

Here $R = 1.5 \times 10^{13}$cm

$r = 7 \times 10^{10}$cm

$S = 2.3$ cal/cm²/minute

$$E = \left[\frac{1.5\times10^{13}}{7\times10^{10}}\right]^2 \times \frac{2.3}{60} \text{ cal/s} \quad ...(i)$$

$$1.37 \times 10^{-12} \times T^4 = \left[\frac{1.5\times10^{13}}{7\times10^{10}}\right]^2 \times \frac{2.3}{60}$$

$$E = \sigma T^4 \quad ...(ii)$$

Equating (i) and (ii)

$$\mathbf{T = 5987K.}$$

Example 8:

In an Ingen-hausz experiment, wax melted over 10 cm of copper rod and over 4 cm of iron rod. What is the conductivity of iron when the conductivity of copper is 0.90?

Solution:

Here, $l_1 = 10\text{cm}$

$l_2 = 4\text{cm}$

$K_1 = 0.90$

$K_2 = ?$

$$\frac{K_2}{K_1} = \frac{l_2^2}{l_1^2}$$

$$K_2 = \frac{l_2^2}{l_1^2} \times K_1$$

$$= \frac{16}{100} \times 0.90$$

$$\mathbf{K_2 = 0.144.}$$

Example 9:

The opposite faces of a metal plate of 0.2 cm thickness an at a difference of temperature of 109°C and the area of the plate is 200 sq cm. Find the quantity of heat that will flow through the plate in one minute if K = 0.2 CGS units.

Solution:

Here,
$$K = 0.2$$
$$A = 200 \text{ sq cm}$$
$$d = 0.2 \text{ cm}$$
$$(\theta_1 - \theta_2) = 100^{\circ}C$$
$$t = 60s$$
$$Q = \frac{KA(\theta_1 - \theta_2)t}{d}$$
$$= \frac{0.2 \times 200 \times 100 \times 60}{0.2}$$
$$= \mathbf{12 \times 10^5 \text{ cal.}}$$

Example 10:

A bar of length 30 cm and uniform area of cross-section 5 cm² consists of two halves AB of copper and BC of iron welded together at B. The end A is maintained at 200°C and the end C at 0°C. The sides of the bar are thermally insulated. Find the rate of flow of heat along the bar when the steady state is reached. Thermal conductivity of copper if 0.9 and thermal conductivity of iron is 0.12 CGS units.

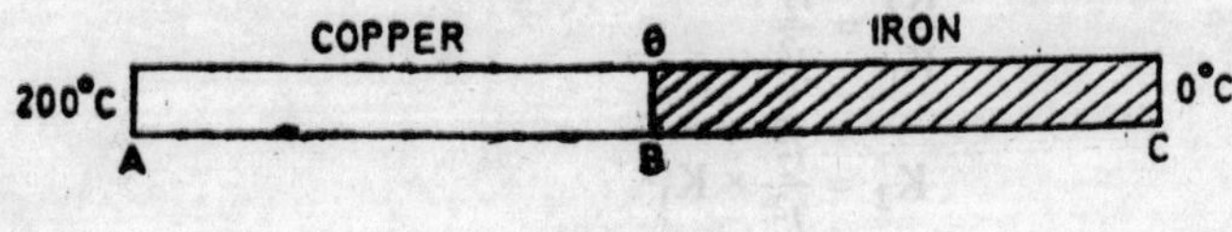

Fig. 3.53

Solution:

Suppose the temperature at the interface. B is θ after the steady state is reached.

$$\therefore \quad \frac{K_1 A(200 - \theta)}{d} = \frac{K_2 A(\theta - 0)}{d}$$
$$0.9(200 - \theta) = 0.12\,\theta$$
$$\theta = 176.5^{\circ}C.$$

After the steady state is reached, the rate of flow of heat is the same in both the bars.

$$\therefore \quad Q = \frac{K_1 A(\theta - 0)}{d}$$

$$Q = \frac{0.9 \times 5 \times (200 - 176.5)}{15}$$

$$\mathbf{Q = 7.05\ cal/s.}$$

Example 11:

An ice box is built of wood 1.76 cm thick, lined inside with cork 3 cm thick. If the temperature of the inner surface of the cork is 0°C and that of the outer surface of wood is 12°C, what is the temperature of the interface ? The thermal conductivity of wood and cork are 0.0006 and 0.00012 CGS units respectively.

Solution:

Suppose the temperature of the interface is 9 after the steady state is reached

$$\therefore \quad \frac{K_1 A(12 - \theta)}{d_1} = \frac{K_2 A(\theta - 0)}{d_2}$$

Here $K_1 = 0.0006$,

$K_2 = 0.00012$

$d_1 = 1.75$ cm,

$d_2 = 3$ cm

$$\frac{0.0006(12 - \theta)}{1.75} = \frac{0.00012(\theta)}{3}$$

$$\mathbf{\theta = 10.74^\circ C.}$$

Example 12:

Find the time in which a layer of ice 3 cm thick on the surface of a pond will increase its thickness by 1 mm when the temperature of the surrounding air is –20°C.

Thermal conductivity of ice = 0.005

Latent heat of ice = 80 cal/g

Density of ice at 0°C = 0.91 g/cm³.

Solution:

$$\int dt = \frac{\rho L}{K\theta} \int x dx$$

$$t = \frac{\rho L}{2K\theta}[x_2^2 - x_1^2]$$

Here, $\rho = 0.91 \text{ g/cm}^3$

$L = 80 \text{ cal/g}$

$K = 0.005$

$\theta = 20^\circ C$

$x_1 = 3 \text{ cm}$

$x_2 = 3.1 \text{ cm}$

$$t = \frac{0.91 \times 80}{2 \times 0.005 \times 20}[(3.1)^2 - (3)^2]$$

$= 222.04 \text{ s}$

= 3 min 42 s approximately.

Example 13:

How much time will it take for a layer of ice of thickness 10 cm to increase by 5 cm on the surface of a pond when the temperature of the surroundings is –10°C?

Solution:

$K = 0.005$

$L = 80 \text{ cal/g}$

$p = 0.90 \text{ g/cm}^3$

Here $\int dt = \frac{\rho L}{K\theta}\int x dx$

$$t = \frac{\rho L}{2K\theta}[x_2^2 - x_1^2]$$

Here $x_1 = 10 \text{ cm},$

$x_2 = 15 \text{ cm}$

$$t = \frac{0.90 \times 80}{2 \times 0.005 \times 10}[(15)^2 - (10)^2]$$

$= \mathbf{9 \times 10^4 s = 25 \text{ hours.}}$

EXERCISES

1. What is Stefan's Law of radiation? Give a laboratory method for determining Stefan's constant.
2. What is meant by black body radiation? State and prove Stefan's law.
3. Describe Planck's few of black body radiation. Describe an experiment to verify the law.
4. Deduce Stefan's law of radiation from the principles of thermodynamics. Describe an experiment to determine Stefan's constant.
5. Give an account of the methods for measuring high temperatures by radiation pyrometry.
6. Describe with necessary theory a method of comparing the conductivities of two metals.
7. Describe Lee's method for determining the thermal conductivity of bad conductors.
8. Define a perfectly black body. Show that an enclosure maintained at a constant temperature behaves as a perfectly black body irrespective of the nature of the walls or the interior. Describe a black body in practice.
9. What is black body radiation? State Stefan's law and describe a method for determining Stefan's constant.
10. Describe the construction and working of a total radiation pyrometer.
11. Describe Lee's method to find the coefficient of thermal conductivity of metals.
12. Discuss the rectilinear flow of heat along a bar of uniform area of cross-section.
13. Discuss Ingen-Hausz experiment to compare the thermal conductivity of different materials.
14. Discuss in detail Forbes method for finding the coefficient of thermal conductivity of a metal bar.
15. Discuss Lee and Chariton's method for finding the coefficient of thermal conductivity for bad conductors.

16. Give the theory of radial flow of heat.
17. Discuss the cylindrical flow of heat. Describe experiments to find the coefficients of thermal conductivity of glass and rubber.
18. Discuss the heat flow through a compound wall.
19. Discuss "accretion of ice" on ponds.
20. Discuss Wiedemann-Franz law.
21. Show how the atmospheric pressure changes with height. Calculate the lapse rate.
22. What do you understand by a black body? State and explain Kirchhoff's laws of heat radiations.
23. State Stefan's law of heat radiation and derive the law mathematically.
24. Describe an experiment for the verification of Stefan's law.
25. Give a laboratory method for determining the Stefan's constant. What is its numerical value in M.K.S. units?
26. Describe the working of:

    (i) disappearing filament pyrometer,

    (ii) total radiation pyrometer, and

    (iii) pyrheliometer.
27. Discuss the distribution of energy in the spectrum of a black body. What are the results obtained by Lummer and Pringsheim?
28. Discuss Wien's displacement law and Rayleigh-Jeans law. How is Planck's law applicable for all wavelengths?
29. Describe the cylindrical shell method of determining the thermal conductivity of poor conductors. Derive the formula you will use.
30. Describe Lee's method for determining the thermal conductivity of poor conductors.
31. Describe Forbes' method for the determination of thermal conductivity of a metal rod. Mention the merits and limitations of this method.
32. State and prove Kirchhoff's law of thermal radiation. Mention one of its illustrative examples.

---

33. State the basic concept on which Planck's law of black-body radiation is based. Write down Planck's formula and show that Wien's and Rayleigh-Jeans laws are particular cases of it.

34. What is quantum theory of radiation? Derive Planck's formula for the distribution of energy in the spectrum of a black body. Deduce from it Wien's displacement law and Rayleigh-Jeans law.

35. Find the time in which a layer of ice 6 cm thick on the surface of a pond will increase its thickness by 2 mm, when the temperature of the surrounding air is –20°C.

    $K = 0.005$, $L = 80$ cal/g, $\rho = 0.91$ g/cm^3

36. How much .time will it take fora layer of ice of thickness 8 cm to increase by 4 cm on the surface of a pond when the temperature of the surroundings is –KFC.

    $K = 0.005$, $L = 80$ cal/g, $\rho = 0.90$ g/cm^3

37. Two large closely spaced concentric spheres (both of which are black body radiators) are maintained at temperatures of 400 K and 600 K respectively. The space in between the two spheres is evacuated. Calculate net rate of energy transfer between the two temperatures.

    $\sigma = 5.672 \times 10^{-8}$ M.K.S. units

38. Calculate the radiant emittance of a black body at a temperature of (i) 200 K and (ii) 2000 K.

    $\sigma = 5.672 \times 10^{-8}$ M.K.S. units

39. The relative emittance of tungsten is approximately 0.70. A tungsten sphere of surface area 5 × 10 4 sq metres is suspended inside a large evacuated enclosure whose walls are at 300 K. What power input is required to maintain the sphere at a temperature of 3000 K. The conduction of heat along the supports can be neglected.

    $\sigma = 5.672 \times 10^{-8}$ M.K.S. units.

40. An aluminium foil of relative emittance 0.2 is placed in between two concentric spheres at temperatures 300 K and 200 K respectively. Calculate the temperature of the foil after the steady state is reached. Assume that the spheres are perfect black body

radiators. Also calculate the rate of energy transfer between one of the spheres and the foil.

$\rho = 5.672 \times 10^{-8}$ M.K.S. units.

41. Calculate the energy radiated per minute from the filament of an incandescent lamp .at 3000 K, if the surface area is 10^{-4} sq metres and its relative emittance is 0-425.

42. An iron furnace radiates 7.65×10^4 calories per hour through an opening of cross section 10^{-4} sq metres. If the relative emittance of the furnace is 0'40, calculate the temperature of the furnace.

43. Write notes on:

    (i) Planck's law

    (ii) Rayleigh-Jean's law

    (iii) Wien's displacement law

    (iv) Convective Equilibrium of the atmosphere

    (v) Optical pyrometer

    (vi) Radiation pyrometer

    (vii) Black body radiation

    (viii) Radiation Pyrometry.

44. Define thermal conductivity and give its units. Describe Lee's method of determining the thermal conductivity of a bad conductor.

45. Explain what you understand by a black body. State Stefan's law of radiation and prove if from thermodynamical considerations. Indicate how it can be verified. How is this law distinguished from Newton's law of cooling?

46. Describe in detail a suitable experimental method of determining the thermal conductivity of a bad conductor.

47. Define coefficient of thermal conductivity. Describe Forbes' method for determining the thermal conductivity of a metal rod. Mention some of the limitations of this method.

48. Prove that at any temperature, the ratio of the emissive power to the absorptive power of a surface is constant and is equal to the emissive power of a perfectly black body.

49. Derive Planck's radiation formula and show that Rayleigh-Jean's law and Wien's law are special cases of Planck's law.

50. State and explain the laws relating to radiation and temperature of a radiating body. What do you mean by:

    (a) perfectly black body,

    (b) coefficient of absorption, and

    (c) coefficient of emission?

51. Define solar constant. How is it experimentally determined? Comment on the source of energy in the sun.

52. Define coefficient of thermal conductivity. What do you understand by a thermal steady state?